计算机专业英语

主　编　刘来毅　杜晓军
副主编　王　莹　韩冬艳　孙晓明　刘慧源

北京理工大学出版社
BEIJING INSTITUTE OF TECHNOLOGY PRESS

版权专有　侵权必究

图书在版编目（CIP）数据

计算机专业英语/刘来毅，杜晓军主编．—北京：北京理工大学出版社，2018.4（2020.8 重印）

ISBN 978-7-5682-5473-1

Ⅰ.①计…　Ⅱ.①刘…②杜…　Ⅲ.①电子计算机 – 英语 – 高等学校 – 教材　Ⅳ.①TP3

中国版本图书馆 CIP 数据核字（2018）第 058774 号

出版发行 / 北京理工大学出版社有限责任公司

社　　址 / 北京市海淀区中关村南大街 5 号

邮　　编 / 100081

电　　话 /（010）68914775（总编室）

　　　　　（010）82562903（教材售后服务热线）

　　　　　（010）68948351（其他图书服务热线）

网　　址 / http：//www.bitpress.com.cn

经　　销 / 全国各地新华书店

印　　刷 / 三河市天利华印刷装订有限公司

开　　本 / 787 毫米 × 1092 毫米　1/16

印　　张 / 10　　　　　　　　　　　　　　　责任编辑 / 梁铜华

字　　数 / 226 千字　　　　　　　　　　　　文案编辑 / 梁铜华

版　　次 / 2018 年 4 月第 1 版　2020 年 8 月第 3 次印刷　责任校对 / 杜　枝

定　　价 / 35.00 元　　　　　　　　　　　　责任印制 / 施胜娟

图书出现印装质量问题，请拨打售后服务热线，本社负责调换

前　　言

随着当今社会的发展，计算机已成为现代科学技术的重要标志，而且由于能够准确无误地对信息进行加工、存储、传送和输出而被人们广泛应用。在这种情况下，计算机专业英语也随之变得尤为重要。它不仅是计算机专业人员的必修课，也是世界上 IT 人员必备的课程，现在已成为每个大中专和本科院校必须开设的课程之一。为此，编者结合教学过程中的经验，编写了本书。

本书是为高职院校计算机相关专业编写的英语教材，主要目的是让学生通过本书的学习，既能掌握一定的专业术语、扩大知识面，又能提高阅读和翻译专业英语文献资料的能力，并通过课堂英语交流，提高学生的英语口语能力，从而能更好地适应信息社会对计算机人才的要求。

本书分为计算机基础、网络和软件三部分，其中计算机基础部分有 7 篇文章，主要包括计算机概述、Windows 简介、计算机发展史、存储器、通信介质、计算机病毒和人工智能 7 方面的内容；网络部分有 6 篇文章，包括局域网、互联网、移动通信、物联网、协议和搜索引擎 6 方面的内容；软件部分有 11 篇文章，包括 Office 365、Photoshop、Adobe Premiere Pro、计算机语言、数据库应用、TurboSquid for 3ds Max、After Effects、Macromedia Dreamweaver、Adobe Fireworks、QuickTime 和 Animate CC 等方面的内容。每篇文章都包括导读、正文、重点词汇、扩展词汇、口语练习、练习题和译文 7 部分。

本书在编写过程中本着简明、易学、实用的原则，语言流畅、通俗易懂、图文并茂。初学者只要对照本书的所讲内容就能很快地掌握计算机专业方面的英语知识。

本书由辽宁农业职业技术学院刘来毅、杜晓军担任主编，王莹、韩冬艳、孙晓明、刘慧源 4 人担任副主编。

本书可作为高等职业院校计算机及相关专业的"计算机专业英语"课程的教材，也可作为广大计算机英语初学者的参考用书。

由于编者水平有限，加上编写时间仓促，书中不足之处在所难免，敬请广大读者批评指正。

目　　录

Chapter 1　Basis of Computer ……………………………………………… 1

  1-1　Computer Overview ………………………………………………… 1

    Introduction（导读）………………………………………………… 1

    Key words and expressions（重点词汇）………………………… 2

    Expanded vocabulary（扩展词汇）………………………………… 2

    Oral practice（口语练习）………………………………………… 2

    Exercises（练习题）………………………………………………… 3

    译文………………………………………………………………… 4

  1-2　Introduction to Windows …………………………………………… 5

    Introduction（导读）………………………………………………… 5

    Key words and expressions（重点词汇）………………………… 6

    Expanded vocabulary（扩展词汇）………………………………… 7

    Oral practice（口语练习）………………………………………… 7

    Exercises（练习题）………………………………………………… 7

    译文………………………………………………………………… 8

  1-3　Introduction to the History of Computers ……………………… 10

    Introduction（导读）………………………………………………… 10

    Key words and expressions（重点词汇）………………………… 11

    Expanded vocabulary（扩展词汇）………………………………… 12

    Oral practice（口语练习）………………………………………… 12

    Exercises（练习题）………………………………………………… 12

    译文………………………………………………………………… 13

  1-4　Memory Devices …………………………………………………… 15

    Introduction（导读）………………………………………………… 15

    Key words and expressions（重点词汇）………………………… 16

    Expanded vocabulary（扩展词汇）………………………………… 17

    Oral practice（口语练习）………………………………………… 17

    Exercises（练习题）………………………………………………… 17

    译文………………………………………………………………… 18

  1-5　Communication Medium …………………………………………… 20

    Introduction（导读）………………………………………………… 20

Key words and expressions（重点词汇） …………………………………… 21
　　Expanded vocabulary（扩展词汇） ……………………………………… 22
　　Oral practice（口语练习） ………………………………………………… 22
　　Exercises（练习题） ……………………………………………………… 22
　　译文 ………………………………………………………………………… 23

1–6　The Computer Virus …………………………………………………… 25
　Introduction（导读） ………………………………………………………… 25
　　Key words and expressions（重点词汇） …………………………………… 27
　　Expanded vocabulary（扩展词汇） ……………………………………… 27
　　Oral practice（口语练习） ………………………………………………… 28
　　Exercises（练习题） ……………………………………………………… 28
　　译文 ………………………………………………………………………… 30

1–7　Artificial Intelligence（AI） ………………………………………… 32
　Introduction（导读） ………………………………………………………… 32
　　Key words and expressions（重点词汇） …………………………………… 33
　　Expanded vocabulary（扩展词汇） ……………………………………… 33
　　Oral practice（口语练习） ………………………………………………… 34
　　Exercises（练习题） ……………………………………………………… 34
　　译文 ………………………………………………………………………… 35

Chapter 2　Network …………………………………………………………… 37

2–1　Local Area Network ……………………………………………………… 37
　Introduction（导读） ………………………………………………………… 37
　　Key words and expressions（重点词汇） …………………………………… 40
　　Expanded vocabulary（扩展词汇） ……………………………………… 40
　　Oral practice（口语练习） ………………………………………………… 41
　　Exercises（练习题） ……………………………………………………… 41
　　译文 ………………………………………………………………………… 43

2–2　Internet …………………………………………………………………… 46
　Introduction（导读） ………………………………………………………… 46
　　Key words and expressions（重点词汇） …………………………………… 47
　　Expanded vocabulary（扩展词汇） ……………………………………… 48
　　Oral practice（口语练习） ………………………………………………… 48
　　Exercises（练习题） ……………………………………………………… 49
　　译文 ………………………………………………………………………… 51

2–3　Mobile Communication ………………………………………………… 52
　Introduction（导读） ………………………………………………………… 52
　　Key words and expressions（重点词汇） …………………………………… 53
　　Expanded vocabulary（扩展词汇） ……………………………………… 53

  Oral practice（口语练习） ……………………………………………………… 53
  Exercises（练习题） ……………………………………………………………… 54
  译文 ………………………………………………………………………………… 55
 2-4 Internet of Things ………………………………………………………………… 57
  Introduction（导读） …………………………………………………………… 57
  Key words and expressions（重点词汇） ………………………………………… 59
  Expanded vocabulary（扩展词汇） ……………………………………………… 59
  Oral practice（口语练习） ……………………………………………………… 59
  Exercises（练习题） ……………………………………………………………… 60
  译文 ………………………………………………………………………………… 61
 2-5 Protocols …………………………………………………………………………… 62
  Introduction（导读） …………………………………………………………… 62
  Key words and expressions（重点词汇） ………………………………………… 63
  Expanded vocabulary（扩展词汇） ……………………………………………… 64
  Oral practice（口语练习） ……………………………………………………… 64
  Exercises（练习题） ……………………………………………………………… 64
  译文 ………………………………………………………………………………… 66
 2-6 Search Engines …………………………………………………………………… 67
  Introduction（导读） …………………………………………………………… 67
  Key words and expressions（重点词汇） ………………………………………… 69
  Expanded vocabulary（扩展词汇） ……………………………………………… 69
  Oral practice（口语练习） ……………………………………………………… 69
  Exercises（练习题） ……………………………………………………………… 70
  译文 ………………………………………………………………………………… 71

Chapter 3 Software ……………………………………………………………………… 72

 3-1 Welcome to the New Office for Home—Office 365 Home ………………… 72
  Introduction（导读） …………………………………………………………… 72
  Key words and expressions（重点词汇） ………………………………………… 73
  Expanded vocabulary（扩展词汇） ……………………………………………… 74
  Oral practice（口语练习） ……………………………………………………… 74
  Exercises（练习题） ……………………………………………………………… 74
  译文 ………………………………………………………………………………… 76
 3-2 What Is Photoshop? ……………………………………………………………… 78
  Introduction（导读） …………………………………………………………… 78
  Key words and expressions（重点词汇） ………………………………………… 79
  Expanded vocabulary（扩展词汇） ……………………………………………… 80
  Oral practice（口语练习） ……………………………………………………… 80
  Exercises（练习题） ……………………………………………………………… 80

| 译文 | 82 |

3-3 What Is Adobe Premiere Pro? … 84
Introduction（导读） … 84
Key words and expressions（重点词汇） … 85
Expanded vocabulary（扩展词汇） … 85
Oral practice（口语练习） … 86
Exercises（练习题） … 86
译文 … 88

3-4 Computer Language—Communication Tool Between Human and Computer … 89
Introduction（导读） … 89
Key words and expressions（重点词汇） … 91
Expanded vocabulary（扩展词汇） … 91
Oral practice（口语练习） … 91
Exercises（练习题） … 92
译文 … 93

3-5 Application Fields of Database … 95
Introduction（导读） … 95
Key words and expressions（重点词汇） … 97
Expanded vocabulary（扩展词汇） … 98
Oral practice（口语练习） … 98
Exercises（练习题） … 99
译文 … 101

3-6 How Can TurboSquid Help You Use 3ds Max? … 104
Introduction（导读） … 104
Key words and expressions（重点词汇） … 105
Expanded vocabulary（扩展词汇） … 105
Oral practice（口语练习） … 105
Exercises（练习题） … 106
译文 … 107

3-7 After Effects CC：What's Changed in This October 2013 Update? … 109
Introduction（导读） … 109
Key words and expressions（重点词汇） … 110
Expanded vocabulary（扩展词汇） … 111
Oral practice（口语练习） … 111
Exercises（练习题） … 111
译文 … 112

3-8 What Is Macromedia Dreamweaver? … 114
Introduction（导读） … 114
Key words and expressions（重点词汇） … 116

  Expanded vocabulary（扩展词汇） ·················· 116
  Oral practice（口语练习） ······················ 117
  Exercises（练习题） ·························· 117
  译文 ······································ 118

3－9 The Future of Adobe Fireworks ·················· 121
  Introduction（导读） ·························· 121
  Key words and expressions（重点词汇） ············ 123
  Expanded vocabulary（扩展词汇） ·················· 124
  Oral practice（口语练习） ······················ 124
  Exercises（练习题） ·························· 125
  译文 ······································ 126

3－10 What Is QuickTime 7？ ························ 129
  Introduction（导读） ·························· 129
  Key words and expressions（重点词汇） ············ 131
  Expanded vocabulary（扩展词汇） ·················· 132
  Oral practice（口语练习） ······················ 133
  Exercises（练习题） ·························· 133
  译文 ······································ 135

3－11 New Features of Animate CC ···················· 137
  Introduction（导读） ·························· 137
  Key words and expressions（重点词汇） ············ 139
  Expanded vocabulary（扩展词汇） ·················· 140
  Oral practice（口语练习） ······················ 140
  Exercises（练习题） ·························· 141
  译文 ······································ 142

Chapter 1

Basis of Computer

1-1 Computer Overview

Introduction (导读)

The application of computers is becoming more common in China. After the reform and opening-up, the number of Chinese computer users continues to rise, and the level of application has been improving continuously, especially in many application fields where good results have been yielded such as the Internet, communication, multimedia, etc.

计算机的应用在中国越来越普遍,改革开放以后,中国计算机用户的数量不断攀升,应用水平不断提高,特别是互联网、通信、多媒体等领域的应用取得了不错的成绩。

Text (文本)

Computer Overview

A computer is a computing machine used for electronic high speed calculation. It has the following functions: numerical calculation, logical calculation and memory-storage. The computer is a modern intelligent electronic device, which can process the massive data automatically and rapidly with the procedures. It consists of hardware system and software system.

The application of computers has penetrated into every field of society, impelling the social development. Scientific Computing: Scientific computing is the earliest application. The computer has the ability of high-speed operation, large storage capacity and continuous operation, by which we can solve various scientific calculation problems which cannot be solved only by people. Translation: Machine translation eliminates the estrangement between different characters and languages. Multimedia: Integrating text, audio, video, animation, graphics and images and other media has become a new thing named multimedia. Computer

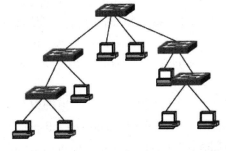

LAN

Network: We overcome the obstacle between the time and space in communication by computer network.

There are many market needs due to the powerful function of computer, so the computer should realize miniaturization and networking, becoming intelligent and giant in the future.

Key words and expressions（重点词汇）

Internet ['ɪntənet] n. 因特网；互联网
network ['netwɜːk] n. 网络 adj. 网络的
data ['deɪtə] n. 数据 adj. 数据的
software ['sɒftweə(r)] n. 软件 adj. 软件的
hardware ['hɑːdweə(r)] n. 硬件 adj. 硬件的
system ['sɪstəm] n. 系统
image ['ɪmɪdʒ] n. 图像
text [tekst] n. 文本；文字
communication [kəˌmjuːnɪ'keɪʃn] n. 通信
video ['vɪdiəʊ] adj. 视频的 n. 录像

Expanded vocabulary（扩展词汇）

multimedia [ˌmʌlti'miːdiə] n. 多媒体 adj. 多媒体的
graphics ['ɡræfɪks] n. 图形；图案
application [ˌæplɪ'keɪʃn] n. 应用
calculation [ˌkælkju'leɪʃn] n. 计算
logical ['lɒdʒɪkl] adj. 逻辑的
memory ['meməri] n. 记忆；内存；存储器
storage ['stɔːrɪdʒ] n. 存储
electronic [ɪˌlek'trɒnɪk] n. 电子 adj. 电子的
process ['prəʊses] n. 过程 vt. 处理
procedures [prə'siːdʒəz] n. 程序
operation [ˌɒpə'reɪʃn] n. 操作；运行
capacity [kə'pæsəti] n. 容量
eliminate [ɪ'lɪmɪneɪt] vt. 消除；去除
integrate ['ɪntɪɡreɪt] vt. 使结合；使一体化
audio ['ɔːdiəʊ] n. 音频；声卡
animation [ˌænɪ'meɪʃn] n. 动画 adj. 动画的

Oral practice（口语练习）

Bill: Hi, Ellen! Could you tell me what a computer is?
Ellen: The computer is a modern intelligent electronic device, which can process the massive data automatically and rapidly with the procedures.

Chapter 1 Basis of Computer

Bill: What can we do using the computer?

Ellen: We can compute datum, translate something, communicate with each other, play multimedia and so on. But can you give me an example about the application of computers?

Bill: Of course! The PPT used by teachers in class is an application of multimedia.

Ellen: Yeah! It's good!

Bill: Thank you for communicating with me!

Ellen: You are welcome!

Exercises (练习题)

Ⅰ. Choices.

1. QQ is _____.
   A. hardware　　　B. software　　　C. audio　　　D. network
2. Which application of computers can help us, if we want to communicate with foreigners?
   A. translation　　B. computing　　C. multimedia　D. others
3. _____ can help us solve various scientific calculation problems.
   A. Multimedia　　B. Network　　　C. Computing　　D. Hardware
4. The _____ is output device.
   A. keyboard　　　B. mouse　　　　C. screen　　　D. disk
5. We can input the words by typing the _____.
   A. keyboard　　　B. mouse　　　　C. screen　　　D. disk

Ⅱ. Translate the following sentences into Chinese.

1. The computer can run rapidly.

2. What software do you know?

3. The data can be transmitted from your telephone to the computer by data lines.

4. Can you set a password for the computer?

5. What applications do you know besides what we have learned?

Ⅲ. Translate the following sentences into English.

1. 请保存好这些文件再关机!

2. 我的计算机是四核的。

3. 我们应当对这台计算机的配置有一个全面的了解。

4. 计算机的应用真的很广泛。

5. 请重新启动那台电脑!

计算机概述

  计算机（computer）俗称电脑，是一种用于高速计算的电子计算机器，既可以进行数值计算，又可以进行逻辑计算，还具有存储记忆功能；它是能够按照程序运行，自动、高速处理海量数据的现代化智能电子设备，由硬件系统和软件系统所组成。
  计算机的应用已渗透到社会的各个领域，推动着社会的发展。科学计算：科学计算是计算机最早的应用领域。利用计算机运算速度高、存储容量大和连续运算的特点，可以解决人工无法完成的各种科学计算问题。翻译：机译消除了不同文字和语言间的隔阂。多媒体：人们把文本、音频、视频、动画、图形和图像等各种媒体综合起来，构成一种全新的概念——"多媒体"（Multimedia）。网络：计算机在网络方面的应用使人类之间的交流跨越了时间和空间障碍。
  计算机强大的应用功能，产生了巨大的市场需要，未来计算机的性能应向着微型化、网络化、智能化和巨型化的方向发展。

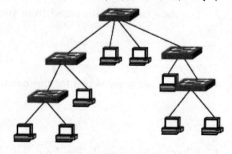

局域网

Chapter 1

Basis of Computer

1-2 Introduction to Windows

Introduction（导读）

Microsoft Windows is a desktop operating system which was developed by Microsoft Corporation in 1985. At first it was only a Microsoft-DOS simulation environment. Microsoft constantly update and upgrade the following systems of the version, and slowly, windows becomes every family's favorite operating system.

Microsoft Windows 是微软公司研发的一套桌面操作系统，问世于 1985 年。它起初仅仅是 Microsoft-DOS 模拟环境。微软公司不断地更新升级此版本的后续系统才使它慢慢地成为家家户户最喜爱的操作系统。

Text（文本）

Introduction to Windows

Microsoft Windows is a software system that works hand in hand with MSDOS to create what is commonly called a graphical operating environment, or Graphical User Interface (GUI). Working with your computer in this environment offers many advantages over working with "normal" MSDOS. Here are some of those advantages:

Windows replaces the DOS command line, so that you no longer have to deal with difficult-to-remember DOS commands. Instead of typing at the DOS A or C prompt to start programs, for example, you can start them by selecting easily recognizable graphic symbols, called icons. And instead of having to look up command syntax in a DOS manual when you want to copy files or check the amount of free space on a disk, you can perform these functions with Windows drop down of menus and dialog boxes. These menus and dialog boxes free you from concerning about command syntax.

Windows lets you run more than one program each time and move easily and quickly between programs. You don't have to quit your word processor, for example, when you want to consult your calendar.

Windows 8

Windows provides a standard mechanism for copying or moving information from one program to another. This mechanism, called the Clipboard, means that information created in one context is instantly reusable in another; you don't need to reenter information or work with clumsy data-transfer utilities.

Windows also includes a facility called Dynamic Data Exchange (DDE) that allows certain program to exchange information automatically. Programs that support DDE can be "hot linked" together so that changes within one are instantly reflected in an other.

Windows makes more efficient use of all your computer's memory than standard MSDOS does. All programs written for version 3 (or later) of Windows can take advantage of memory beyond 640KB, without requiring special hardware or "driver" files.

Windows encourages consistency among applications, making it easier for you to use a variety of complementary programs in your work. Applications written specifically for the Windows environment have a great deal in common, because they all use standard Windows drop-down menu and dialog box formats. Thus, when you learn to use one Windows program, you're well on your way toward knowing how to use a Windows program.

In short, Windows is a system that makes your computer easier to use, allows you to be more productive and gets the maximum value from your hardware and software investment.

Key words and expressions（重点词汇）

graphical ['græfɪkl] *adj.* 绘成图画似的；绘画的；图形化
interface ['ɪntəfeɪs] *n.* 界面；<计>接口；交界面
icon ['aɪkɒn] *n.* 图标；图示；图像；偶像
function ['fʌŋkʃn] *n.* 功能；作用；应变量；函数；职务；重大聚会；
　　　　　　　　　　　 功能；函数；职能；作用
Menu ['menjuː] *n.* 菜单；菜；菜肴；选单；菜单键
syntax ['sɪntæks] *n.* 语法；句法；有秩序的排列

Chapter 1 Basis of Computer

mechanism ['mekənɪzəm] *n.* 机制；机构；机理；机械装置
Dynamic [daɪ'næmɪk] *adj.* 动态的；动力的；动力学的；充满活力的；
  精力充沛的；不断变化的；充满变数的
version ['vɜːʃn] *n.* 版本；译文；译本；说法；倒转术
consistency [kən'sɪstənsi] *n.* 一致性；稠度；一贯性；连贯性

Expanded vocabulary（扩展词汇）

operate ['ɒpəreɪt] *vt.* 操作；控制；使运行
simulation [ˌsɪmju'leɪʃn] *n.* 模仿；模拟
environment [ɪn'vaɪrənmənt] *n.* 环境；外界
favorite ['feɪvərɪt] *adj.* 喜爱的；宠爱的；中意的
advantage [əd'vɑːntɪdʒ] *n.* 优势；优点；好处；有利条件
recognize ['rekəgnaɪz] *vt.* 认出；识别；承认
dialog ['daɪəlɒg] *n.* 对话框；对白；对话；会话
command [kə'mɑːnd] *n.* 命令；指挥
calendar ['kælɪndə(r)] *n.* 日历；历法；日程表

Oral practice（口语练习）

1. Notebooks are expected to run better on Windows 7 than Vista, which required more powerful hardware than notebooks offered.
2. Different operating systems address these issues in different manners, however most operating systems contain components which have similar functionalities.
3. The techniques used to implement these functions may vary from one OS to another, but the fundamental concepts are the same.

Exercises（练习题）

Ⅰ. Choices.

1. _____ is not an OS.
   A. Windows 7   B. Word   C. Linux   D. Unix
2. Windows replaces the _____ command line.
   A. DOS   B. Windows 7   C. Vista   D. Windows 2000
3. Windows is a _____ that makes your computer easier to use.
   A. hardware   B. software   C. system   D. program
4. What does Windows include?
   A. DDE   B. DDR.   C. ROM   D. RAM
5. _____ is a desktop operating system which was developed by Microsoft Corporation in 1985.
   A. DOS   B. Unix   C. Vista   D. Microsoft Windows

Ⅱ. Translate the following sentences into Chinese.

Operating system (OS) functions have evolved in response to the following considerations and issues:

1. Efficient utilization of computing resources.

2. New features in computer architecture.

3. New user requirements.

Ⅲ. Translate the following paragraph into English.

操作系统的基本功能是在计算机系统中组织用户计算的执行，因此它在设计中要求对用户计算有恰当的特征描述，通常在用户计算中使用以下三种特征记述——作业、程序和进程。

Windows 简介

Microsoft Windows 是一个软件系统，它与 MSDOS 联合创立了一个我们平常所说的图形操作环境或者是图形用户接口（GUI）。在这种环境下工作的计算机具有了比"普通"工作在 MSDOS 下的计算机更突出的优点。现列举其中一些：

Windows 取代了 DOS 命令行，因此你不必再记忆那些难以记忆的 DOS 命令。举个例子，你能够通过选择容易识别的图形符号（称为图标）来启动程序，淘汰了原来在 DOS 下的 A 或者 C 提示符下的启动。你也不必为了复制文件或检查磁盘空间而去查看 DOS 手册中的命令语法。在 Windows 中这些功能的执行是通过下拉菜单和对话框来实现。这些菜单和对话框使你从命令语法中解放出来。

Windows 可以同时运行多个程序，并且可简单、迅速地在程序间切换。例如，当你想查看日程时，你不必退出在运行的字处理程序再去操作。

Windows 提供了一套标准的机制，使人们能够在程序之间复制或移动信息。这种

Chapter 1 Basis of Computer

Windows 8

机制称为剪贴板。其意思是说，在一个环境中创建的信息可以立即在另一个环境中使用，你不必重新输入信息或使用笨拙的数据传送功能。

Windows 还包含了一个称作动态数据交换（DDE）的工具，它允许某些程序自动进行信息交换。支持动态数据交换（DDE）的程序能被紧密地链接在一起，从而在一个程序中所做的变更会立即"映射"到其他程序。

Windows 使用计算机内存的效率比标准的 MSDOS 要高。所有 Windows 3（或更高）的程序能够使用超过 640KB 的内存，而不需要特殊的硬件或"驱动"文件。

Windows 一贯支持众多应用程序的一致性，使在工作中能够很容易地使用各种不同的功能互补的程序。专为在 Windows 环境下运行而写的应用程序之所以有不少操作是共同的，是因为使用了标准的 Windows 的下拉菜单和对话框格式。这样，当学会使用一个 Windows 程序时，你就已经完全学会了如何使用其他任何一个 Windows 程序了。

总之，Windows 是一个使计算机更容易使用的系统，它能使你的效率更高，并且从软硬件的投资上得到更多的收益。

Chapter 1

Basis of Computer

1-3 Introduction to the History of Computers

Introduction（导读）

It is hard to say exactly when the modern computer was invented. Starting in the 1930s and through the 1940s, a number of machines were developed that were like computers. But most of these machines did not have all the characteristics that we associate with computers today. These characteristics are that the machine is electronic, that it has a stored program, and that it is general-purpose.

很难确切地说现代计算机是什么时候发明的。20 世纪 30—40 年代，许多类似计算机的机器被开发出来。但是这些机器大部分没有今天我们所说的计算机的所有特征。这些特性是电子设备、具有储存的程序、具有通用性。

Text（文本）

Introduction to the History of Computers

One of the first computerlike devices was developed in Germany in 1941. Called the Z3, it was a general-purpose, stored-program machine with many electronic parts, but it had a mechanical memory. Another electromechanical computing machine was developed by Howard Aiken, with financial assistance from IBM, at Harvard University in 1943. It was called the Automatic Sequence Control Calculator Mark Ⅰ, or simply the Harvard Mark Ⅰ. Neither of these machines was a true computer, however, because they were not entirely electronic.

Perhaps the most influential of the early computerlike devices was the Electronic Numerical Integrator and Computer, or ENIAC. It was developed by J. Presper Eckert and John Mauchly at the University of Pennsylvania. The project began in 1943 and was completed in 1946. The machine was huge; it weighed 30 tons and contained over 18,000 vacuum tubes.

The ENIAC was a major advancement for its time. It was the first general-purpose, electronic computing machine and was capable of performing thousands of operations per second. It was controlled, however, by switches and plugs that had to be manually set. Thus, although it was a

Chapter 1　Basis of Computer

general-purpose electronic device, it did not have a stored program. Therefore, it did not have all the characteristics of a computer.

First-generation computers were characterized by the use of vacuum tubes as their principal electronic component. Vacuum tubes are bulky and produce a lot of heat, so first-generation computers were large and required extensive air conditioning to keep them cool. In addition, because vacuum tubes do not operate very fast, these computers were relatively slow.

In the second generation of computers, transistors replaced vacuum tubes. Although invented in 1948, the first all-transistor computer did not become available until 1959. Transistors are smaller and less expensive than vacuum tubes, and they operate faster and produce less heat. Hence, with second-generation computers, the size and cost of computers decreased; their speed increased, and their air-conditioning needs were reduced.

The technical development that marks the third generation of computers is the use of integrated circuits or ICs in computers. An integrated circuit is a piece of silicon (a chip) containing numerous transistors. One IC replaces many transistors in a computer, resulting in a continuation of the trends beginning in the second generation. These trends include reduced size, reduced cost, increased speed, and reduced need for air conditioning.

Although not everyone agrees that there is a fourth computer generation, those that do feel that it began in 1971, when IBM introduced its successors to the System/360 line of computers. These mainframe computers were called the System/370, and current-model IBM computers, although not called System/370s, evolved directly from these computers.

Software development during the fourth computer generation started off with little change from the third generation. Operating systems were gradually improved, and new languages were designed. Database software became widely used during this time. The most important trend, however, resulted from the microcomputer revolution. Packaged software became widely available for microcomputers so that today most software is purchased, not developed from scratch.

We may have defined our last generation of computers and begun the era of generationless computers. Even though computer manufacturers talk of "fifth-" and "sixth-" generation computers, this talk is more a marketing play than a reflection of reality.

Key words and expressions（重点词汇）

associate [əˈsəʊʃɪət] vt. 联想
general-purpose adj. 通用的
assistance [əˈsɪst(ə)ns] n. 援助；帮助
influential [ˌɪnfluˈenʃl] adj. 有影响的
advancement [ədˈvɑːnsmənt] n. 先进
thousands of 成千上万的
manually [ˈmænjʊəlɪ] adv. 人工地
bulky [ˈbʌlki] adj. 笨重的；庞大的；体积大的
transistor [trænˈzɪstə(r)] n. 晶体管

purchase ['pɜːtʃəs] v. 购买；采购
from scratch 从头做起，从零开始
reflection [rɪ'flekʃn] n. 反映

Expanded vocabulary（扩展词汇）

computerlike [kəm'pjuːtə(r)laɪk] adj. 计算机似的
result in 导致；造成……结果
electromechanical [ɪˌlektrəʊmɪ'kænɪk(ə)l] adj. 机电的；电机的
integrated circuits n. 集成电路
Pennsylvania [ˌpensɪl'veɪnjə] n. 宾夕法尼亚州（美国州名）
vacuum tube 真空管
air conditioning 空气调节
mainframe ['meɪnˌfreɪm] n. 主机；大型机

Oral practice（口语练习）

1. The UNIVAC Ⅰ was the first commercial computer in this generation. As noted earlier, it was used in the Census Bureau in 1951. It was also the first computer to be used in a business application. In 1954, General Electric took delivery of a UNIVAC Ⅰ and used it for some of its business data processing.

2. Software also continued to develop during this time. Many new programming languages were designed, including COBOL in 1960. More and more businesses and organizations were beginning to use computers for their data processing needs.

3. The fourth generation of computers is more difficult to define than the other three generations. This generation is characterized by more and more transistors being contained on a silicon chip. First there was Large Scale Integration (LSI), with hundreds and thousands of transistors per chip, then came Very Large Scale Integration (VLSI), with tens of thousands and hundreds of thousands of transistors. The trend continues today.

Exercises（练习题）

Ⅰ. Choices.

1. How many generations of computers are introduced in this chapter?
   A. 3   B. 4   C. 5   D. 6

2. The first computer's name is _____.
   A. Mark Ⅰ   B. IBM   C. ENIAC   D. 6 System/360

3. The first-generation computers were characterized by the use of _____.
   A. integrated circuits   B. transistors
   C. electronic machines   D. vacuum tubes

4. The second-generation computers were characterized by the use of _____.

Chapter 1 Basis of Computer

    A. integrated circuits   B. transistors

    C. electronic machines   D. vacuum tubes

5. The third-generation computers were characterized by the use of _____.

    A. integrated circuits   B. transistors

    C. electronic machines   D. vacuum tubes

Ⅱ. Translate the following paragraph into Chinese.

Advocates of the concept of generationless computers say that even though technological innovations are coming in rapid succession, no single innovation is, or will be, significant enough to characterize another generation of computers.

Ⅲ. Translate the following paragraph into English.

许多新的编程语言被发明,包括1960年发明的COBOL。越来越多的企业和组织开始使用计算机以满足他们的数据处理需要。

计算机发展史

    第一个类似计算机的装置之一是1941年由德国研制的,叫Z3,它是通用型储存程序机器,具有许多电子部件,但是它的存储器是机械的。另一台机电式计算机器是由霍华德·艾坎在IBM的资助下于1943年在哈佛大学研制的。它被称为自动序列控制计算器Mark Ⅰ,或简称哈佛Mark Ⅰ。然而,这些机器都不是真正的计算机,因为它们不是完全电子化的。

    也许早期最具影响力的类似计算机的装置应该是电子数字积分计算机,或简称ENIAC。它是由宾夕法尼亚大学的J. Presper Eckert和John Mauchly研制的。该工程于1943年开始,并于1946年完成。这台机器极其庞大,重达30吨,而且包含18 000多个真空管。

    ENIAC是当时重要的成就。它是第一台通用型电子计算机器,并能够执行每秒数千次运算。然而,它是由开关和继电器控制的,必须手工设定。因此,虽然它是一个通用型电子装置,但是它没有储存程序。因此,它不具备计算机的所有特征。

    第一代计算机的特色是使用真空管为其主要电子器件。真空管体积大且发热严重,因此第一代计算机体积庞大,并且需要大量的空调设备保持冷却。此外,因为真空管运行不是很快,这些计算机运行速度相对较慢。

    在第二代计算机中,晶体管取代了真空管。虽然发明于1948年,但第一台全晶体管计算机直到1959年才成为现实。晶体管比真空管体积小、价格低,而且运行快且发热少。因此,随着第二代计算机的出现,计算机的体积和成本降低、速度提高,且它们对空调的需要减少。

作为第三代计算机标志的技术发展是在计算机中使用集成电路或简称IC。一个集成电路就是包含许多晶体管的一个硅片（芯片）。一个集成电路代替了计算机中的许多晶体管，从而导致始于第二代的一些趋势继续存在。这些趋势包括计算机体积减小、成本降低、速度提高和对空调的需要减小。

虽然并不是每个人都同意存在一个第四代，那些认为存在的觉得它开始于1971年，其时IBM开发了System/360系列计算机的下一系列产品。这些大型计算机称为System/370，当前的IBM计算机虽然不叫作System/370，但都是从这些计算机直接发展而来的。

在计算机的第四代期间，软件的发展开始与第三代有所不同。操作系统在逐渐地改进，而新的语言被发明。期间数据库软件被广泛使用。然而，最重要的趋势起因于微型计算机革命。用于微型计算机的软件包随处可得，因此今天大多数的软件可以购得，而不需从头开始开发。

我们可能已经定义了我们最新一代计算机而且开始了计算机的无代时代。即使计算机制造商谈到"第五"和"第六"代计算机，这些说法更多是市场行为，而不是真实的反映。

Chapter 1

Basis of Computer

1-4 Memory Devices

Introduction (导读)

In this section, we will discuss the structure and function of the memory subsystem in the computer. We will review the internal composition of different types of physical memory and its chips. We will discuss the construction of the memory subsystem.

本节我们将讨论计算机中存储器子系统的结构和功能。我们将回顾不同类型的物理存储器及其芯片的内部组成，讨论存储器子系统的结构。

Text (文本)

Memory Devices

1. Random-access memory

Random-access memory, or RAM, is the kind of memory we usually refer to when we speak of computer memory. It is the most widely used type, and consists of rows of chips with locations established in tables maintained by the control unit.

The memory

As the name suggests, items stored in RAM can be gotten (accessed) both easily and in any order (randomly) rather than in some sequence. RAM relies on electric current for all its operations; moreover, if the power is turned off or interrupted, RAM quickly empties itself of all your hard work. Thus, we say RAM is volatile, or nonpermanent.

2. Read-only memory

Read-only memory, or ROM, typically holds programs. These programs are manufactured, or "hard-wired" in place on the ROM chips. For example, a microcomputer has a built-in ROM chip (sometimes called ROM BIOS, for ROM basic input/output system) that stores critical programs such as the one that starts up, or "boots," the computer. ROM is "slower" than RAM memory, and as a result, items in ROM are transferred to RAM when needed for fast processing.

Items held in ROM can be read, but they cannot be changed or erased by normal input methods. New items cannot be written into ROM. The only way to change items in most forms of ROM is to change the actual circuits.

3. Magnetic disks

The magnetic disk is a circular platter with a smooth surface and a coating that can be magnetized. Data is stored on it as magnetized spots. The reading and recording device, the disk drive, spins the disk past read/write heads that detect or write the magnetized spots on the disk.

The magnetic disk

4. CD-ROMs

Optical disks need thin beams of concentrated light to store and read data. It is a form of laser storage, called CD-ROM.

There are two types of optical disks that can be user-recorded: WORM and erasable optical. WORM stands for "write once, read many": Data can be written to this disk just one time, but the data can be read many times. Erasable optical disks can be written to, read, and erased.

5. Magnetic tape

A magnetic tape is a narrow plastic strip similar to the tape used in tape recorders. The tapes are read by tape drive moving the tape past a read/write head, which detects or writes magnetized spots on the iron-oxide coating of the tape. Each pattern of spots matches the byte code for character being stored.

The magnetic tape

Key words and expressions（重点词汇）

random ['rændəm] *adj.* [数] 随机的；任意的；胡乱的

chip [tʃɪp] *n.* 芯片；切屑；碎片；薄片

Chapter 1 Basis of Computer

establish [ɪ'stæblɪʃ] vt. 建立；确立；设立
manufacture [mænjʊ'fæktʃə] vt. 制造；加工
boot [buːt] vt. [计算机科学] 引导；启动
processing [prəʊ'sesɪŋ] n. （数据）处理；整理；配置
circuits ['səːkɪt] n. 电路；线路
magnetic [mæg'netɪk] adj. 有磁性的；有吸引力的
detect [dɪ'tekt] vt. 查明；发现；[电子学] 检波
optical ['ɒptɪkl] adj. 视觉的；视力的；眼睛的；光学的

Expanded vocabulary（扩展词汇）

access ['ækses] vt. 使用；存取；接近
refer to 参考；涉及；指的是；适用于
consists of 包含；由……组成；充斥着
unit ['juːnɪt] n. 单位；单元；装置
item ['aɪtəm] n. 条款；项目；一则
smooth [smuːð] adj. 顺利的；光滑的；平稳的
beam [biːm] n. 梁；栋梁；束；光线
erasable [ɪ'reɪsəbl] adj. 可消除的；可抹去的
plastic ['plæstɪk] n. 塑料制品

Oral practice（口语练习）

1. ROM and RAM are very important to the OS. Part of a computer's operating system is built into ROM. That part contains the most essential programs that the computer needs in order to run correctly.
2. The ROM operating system is also known as the BIOS (Basic Input Output System) which is responsible for waking up the computer when you turn it on to remind it of all the parts it has and what they do.
3. The part of the operating system that contains these programs is stored on a computer's hard drive and is booted to RAM whenever the computer is turned on.

Exercises（练习题）

Ⅰ. Choices.

1. Which kind of memory is nonpermanent?
   A. Rom.   B. CD-ROMs.   C. Ram.   D. Magnetic disk.
2. The _____ is a circular platter with a smooth surface and a coating that can be magnetized.
   A. Rom   B. CD-ROMs   C. Ram   D. magnetic disk
3. Erasable optical disks can not be _____.
   A. written   B. read   C. erased   D. programed

4. A magnetic tape is a narrow plastic strip similar to the tape used in _____.
   A. tape recorders   B. Rom   C. the disk   D. Ram
5. _____ is a circular platter with a smooth surface and a coating that can be magnetized.
   A. System   B. The magnetic disk
   C. The disk   D. Optical disk

Ⅱ. Translate the following paragraph into Chinese.

As the number of locations increases, the size of the address decoder needed in a linear organization becomes prohibitively large. To remedy this problem, the memory chip can be designed using multiple dimensions of decoding.

Ⅲ. Translate the following paragraph into English.

构造包含一个简单芯片的存储器是非常容易的，我们只需要简单地从系统总线上连接地址信号线、数据信号线和控制信号线就完成了。然而，大多数的存储器系统需要多个芯片。下面是通过存储器芯片组合来形成存储器子系统的一些方法。

存 储 器

1. 随机存储器

随机存储器或称RAM，是人们常说的计算机存储器。它是使用最广泛的一种类型，由一组芯片构成，其存储单元由控制部件中的地址表管理。

顾名思义，RAM中的内容可以很容易地以任意顺序（随机地）读出（或称访问），而不用考虑什么先来后到。RAM始终靠电流进行操作；而且，如果电源关闭或中断，RAM会很快丢失人们辛辛苦苦存入的内容。因此人们把RAM称为易失性或非永久性的存储器。

内存

2. 只读存储器

只读存储器或ROM，主要用来存储程序。这些程序是做在ROM芯片中或者说通过硬连线实现的。例如在微型计算机中就有内置ROM芯片（有时称为ROM BIOS，即ROM基本输入输出系统），该芯片存储一些关键的程序，如计算机的启动或引导程序。ROM 的读取速度比 RAM 慢，因此如果需要快速处理，则将 ROM 中的内容传送到 RAM 中。

ROM 中的内容可以读，但不能用一般的输入方法更改或擦除。新的内容也不能写入 ROM。在大多数型号 ROM 中变更内容的唯一方法是更换实际电路。

3. 磁盘

磁盘是一个表面光滑且涂敷可磁化材料的圆盘。数据在磁盘上作为磁化点存入。读出与记录设备即磁盘驱动器驱动磁盘转动，当通过读写磁头时，磁头从磁盘上的磁化点检测出或写入信息。

硬盘

4. 只读光盘

光盘用细聚光光束去存储和读出数据。CD-ROM 是激光存储器的一种形式。

用户使用的可读写光盘有两种类型：WORM 和可擦除的光盘。WORM 代表一次写多次读：数据只能写入一次但可读出多次。可擦除光盘可以读、写和擦除。

5. 磁带

磁带是一种类似于录音磁带的窄塑料带。由磁带驱动器去读磁带的内容。驱动器驱动磁带，由读写磁头检出或写入由氧化铁涂敷的磁化点，以各磁化点的不同极性表示所存储字符的二进制代码。

磁带

Chapter 1

Basis of Computer

1-5　Communication Medium

Introduction（导读）

There are two kinds of medium to connect different devices. They are physical connections and wireless connections. We will introduce them in this text.

连接不同设备的通信介质分为两大种类，分别是物理连接和无线连接。我们将在这篇课文里介绍它们。

Text（文本）

Communication Medium

Physical connections use a solid medium to connect sending and receiving devices. These connections include telephone lines (twisted pair), coaxial cable, and fiber-optic cable.

Telephone lines you see strung on poles consist of twisted-pair cable, which is made up of hundreds of copper wires. A single twisted pair culminates in a wall jack into which you can plug your phone and computer. Telephone lines have been the standard transmission medium for years for both voice and data. However, they are now being phased out by more technically advanced and reliable media. Coaxial cable, a high-frequency transmission cable, replaces the multiple wires of telephone lines with a single solid-copper core. In terms of the number of telephone connections, a coaxial cable has over 80 times the transmission capacity of twisted pair. Coaxial cable is used to deliver television signals as well as to connect computers in a network.

Fiber-optic cable transmits data as pulses of light through tiny tubes of glass. In terms of the number of telephone connections, fiber-optic cable has over 26,000 times the transmission capacity of twisted-pair cable. Compared to coaxial cable, it is lighter and more reliable at transmitting data. It transmits information using beams of light at light speeds instead of pulses of electricity, making it far faster than copper cable. Fiber-optic cable is rapidly replacing twisted-pair cable telephone lines.

Wireless connections do not use a solid substance to connect sending and receiving devices. They use the air itself. Primary technologies used for wireless connections are infrared, broadcast radio, microwave, and satellite.

Infrared uses infrared light waves to communicate over short distances. It is sometimes referred to as line-of-sight communication because the light waves can only travel in a straight line. This requires that sending and receiving devices must be in clear view of one another without any obstructions blocking that view. One of the most common applications is to transfer data and information from a portable device such as a notebook computer or PDA to a desktop computer.

Broadcast radio uses radio signals to communicate with wireless devices. For example, cellular telephones and many Web-enabled devices use broadcast radio to place telephone calls and/or to connect to the Internet. Some end users connect their notebook or handheld computers to a cellular telephone to access the Web from remote locations. Most of these Web-enabled devices follow a standard known as WI-FI (wireless fidelity). This wireless standard is widely used to connect computers to each other and to the Internet.

Microwave communication uses high-frequency radio waves. Like infrared, microwave communication provides line-of-sight communication because microwaves travel in a straight line. Because the waves cannot bend with the curvature of the earth, they can be transmitted only over relatively short distances. Thus, microwave is a good medium for sending data between buildings in a city or on a large college campus. For longer distances, the waves must be relayed by means of microwave stations. These stations can be installed on towers, high buildings, and mountains.

Bluetooth is a short-range wireless communication standard that uses microwaves to transmit data over short distances of up to approximately 33 feet①. Unlike traditional microwaves, Bluetooth does not require line-of-sight communication. Rather, it uses radio waves that can pass through nearby walls and other nonmetal barriers. It is anticipated that within the next few years, this technology will be widely used to connect a variety of different communication devices.

Key words and expressions（重点词汇）

associate [əˈsoʃɪet] *vt.* 联想
coaxial cable *n.* 同轴电缆
twisted-pair cable 双绞线
reliable [rɪˈlaɪəbl] *adj.* 可靠的
substance [ˈsʌbstəns] *n.* 物质；材料
infrared [ˌɪnfrəˈred] *n.* 红外线
cellular telephone 移动电话
end user 最终用户；终端用户
relay [ˈriːleɪ] *vt.* 转播
approximately [əˈprɔksɪmətli] *adv.* 近似地；大约

---

① 1 feet = 0.304 8 meter.

Expanded vocabulary（扩展词汇）

copper ['kɔpə(r)] *n.* 铜
in terms of 就……而言
capacity [kə'pæsəti] *n.* 容量；性能
line-of-sight 视距；视线范围内
obstruction [əb'strʌkʃn] *n.* 障碍物
portable ['pɔːtəbl] *adj.* 轻便的；手提的
bend [bend] *vt.* 弯曲

Oral practice（口语练习）

1. Satellite communication uses satellites orbitting about 22,000 miles above the earth as microwave relay stations. Many of these are offered by Intelsat, the International Telecommunications Satellite Consortium, which is owned by 114 governments and forms a worldwide communication system.
2. Satellites rotate at a precise point and speed above the earth. They can amplify and relay microwave signals from one transmitter on the ground to another. Satellites can be used to send and receive large volumes of data. Uplink is a term relating to sending data to a satellite. Downlink refers to receiving data from a satellite.
3. One of the most interesting applications of satellite communications is for global positioning. A network of 24 satellites owned and managed by the Defense Department continuously sends location information to the earth.

Exercises（练习题）

Ⅰ. Choices.

1. How many kinds of communication media are introduced in this chapter?
   A. 2.　　　　　　B. 3.　　　　　　C. 6.　　　　　　D. 7.
2. Which is a solid medium to connect sending and receiving devices?
   A. Bluetooth.　　B. Microwave.　　C. Fiber-optic.　　D. Infrared.
3. Which do not require line-of-sight communication?
   A. Bluetooth.　　B. Infrared.　　　C. Microwave.　　D. WI-FI.

Ⅱ. Questions and answers.

1. Please list the other communication media which you know.

2. Say something about the advantages and disadvantages of the communication media which you list.

Ⅲ. Translate the following paragraph into Chinese.

Global positioning system (GPS) devices use that information to uniquely determine the geographical location of the device. Available in some automobiles to provide navigational support, these systems are often mounted into the dash with a monitor to display maps and speakers to provide spoken directions.

Ⅳ. Translate the following paragraph into English.

大量的计算机通信通过电话线传播。然而,电话由于最初是设计用来传输信息的,所以只发送和接收模拟信号。相反,它不发送和接收数字信号。所以想用电话线来上网,你需要一个调制解调器。

通信介质

物理连接使用固体介质连接发送和接收设备。这些连接包括电话线路(双绞线)、同轴电缆和光纤电缆。

你所看到的电话线,是由上百根铜线缠绕而成的双绞线。双绞线的一个端口是固定电话,当然也可以插在手机和电脑上。电话线作为一个标准传输媒介,多年来一直在负责语音和数据的传输。然而,他们现在已经被更先进、可靠的媒体所淘汰了。同轴电缆、高频传输电缆,取代了多线缆的电话线。在传输能力方面,一个同轴电缆是双绞线传输容量的80倍。同轴电缆现在主要用于提供电视信号以及连接网络的电脑。

光纤电缆传输数据是由光脉冲在细小的玻璃导管内进行的。在传输能力上,光纤电缆的传输速度已经超过双绞线26 000倍。与同轴电缆相比,光纤更轻、更可靠。光纤使用光来传输信息,而不是电脉冲,这使他们传输速度远远快于铜电缆。光纤电缆正在迅速取代电缆电话线路。

无线连接不使用固体物质连接发送和接收数据。他们使用空气本身传播。目前此项技术主要用于红外连接、广播无线电、微波和卫星。

红外技术使用红外光波在短距离内通信。它有时被称为视距通信,因为光只能沿直线传播。这就要求发送和接收设备必须没有任何障碍物阻止。最常见的应用是用来传送数据和信息,是从一个便携式设备上,如笔记本电脑或掌上电脑,往桌面计算机上传输数据。

广播电台使用无线电信号与无线设备通信。例如，移动电话和许多网络设备使用广播电台拨打电话和/或连接到互联网。一些终端用户将他们的笔记本电脑或便携式电脑通过手机连接上网。大多数的这些网络设备遵循一个标准，称为WI-FI。这个无线标准被广泛用于局域网和因特网。

微波通信使用高频无线电波。如红外、微波通信只能提供短距离通信因为微波以直线的方式行进。因此，微波是非常适合用来在大城市或者校园内作为数据传输的媒介。微波发射台可以安装在塔或者高的建筑之上。

蓝牙是一种短距离的无线通信标准，通信距离大约33英尺。与传统的微波相比，蓝牙不需要视距通信。相反，它使用无线电波可以穿透附近的墙壁和其他非金属壁垒。在未来几年中，这种技术将被广泛用于连接各种不同的通信设备。

Chapter 1

Basis of Computer

1-6 The Computer Virus

Introduction（导读）

Has your computer been infected with the viruses? Computer viruses are called viruses because they share some of the traits of biological viruses. A computer virus passes from computer to computer like a biological virus which passes from person to person.

你的计算机感染过病毒吗？计算机病毒之所以被称作病毒是因为它们具有生物病毒的共同特点。计算机病毒在计算机与计算机之间的传播很像生物病毒在人与人之间的传播。

Text（文本）

The Computer Virus

1. Computer virus definition

What is a computer virus? All that can cause computer fault, destruction of computer data and program collectively are referred to as a computer virus. The viruses we heard of such as MYDOOM, PHISHING, Gray pigeons are all computer viruses.

A computer virus is a program designed to spread itself by first infecting executable files or the system areas of hard and floppy disks and then making copies of itself. There are many types of viruses: Trojan, Worm, Backdoor, Script, Macro, Win32 PE, Binder, downloader. Common viruses are as follows.

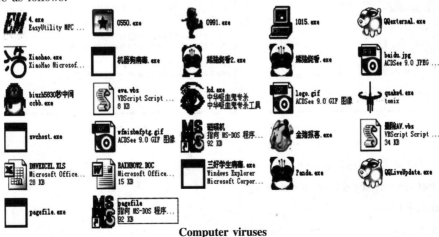

Computer viruses

2. How do viruses spread?

- Over execute program.

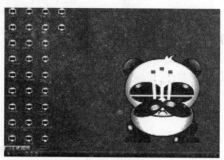

Execute program

- Over a network.

Network

- Over flash memory.

Flash memory

The spread of the virus through the Internet is prompt—only one day!

3. What do viruses do to computers?

- Damage files.
- Interfere with your computer's operations.

- Try to spread themselves around.

Note that viruses can't do any damage to hardware: Do not melt down your CPU; do not burn out your hard drive; do not cause your monitor to explode. Some general tips on avoiding virus infection: To data important files backup; install anti-virus software; scanning for viruses on a regular basis. There are some common antivirus softwares: 360 antivirus, Rising Antivirus, Kingsoft Antivirus.

Last, let me introduce the Hacker: In the computer security context, a hacker is someone who seeks and exploits weaknesses in a computer system or computer network. Hackers may be motivated by a multitude of reasons, such as profit, protest, or challenge. While other uses of the word hacker exist that are not related to computer security, such as referring to someone with an advanced understanding of computers and computer networks, they are rarely used in mainstream context. They may be good at computer, but in the wrong direction. Although we are just programmers, but we have to have a principle.

Key words and expressions（重点词汇）

virus ['vaɪrəs] *n.* 病毒；计算机病毒
hack [hæk] *n.* 黑客
destruction [dɪ'strʌkʃn] *n.* 破坏；毁灭
data ['deɪtə] *n.* 数据；资料
design [dɪ'zaɪn] *vt. & vi.* ; *n.* 设计
infect [ɪn'fekt] *vt.* 使受影响；感染；传染
executable [ɪɡ'zekjuətəbl] *adj.* 实行的；执行的；可执行的
floppy disk 软盘
memory ['meməri] *n.* 记忆［原意］；存储器；内存［计算机］
definition [ˌdefɪ'nɪʃn] *n.* 定义；规定；明确
damage ['dæmɪdʒ] *vt. & vi.* ; *n.* 损害；毁坏
hardware ['hɑːdweə(r)] *n.* 五金器具［原意］；计算机硬件［计算机］
monitor ['mɒnɪtə(r)] *n.* 显示器
backup ['bækʌp] *n.* 备份
software ['sɒftweə] *n.* 软件
antivirus ['æntivaɪrəs] *n.* 杀毒
network ['netwɜːk] *n.* 网络

Expanded vocabulary（扩展词汇）

collectively [kə'lektɪvlɪ] *adv.* 全体地；共同地
prompt [prɒmpt] *adj.* 敏捷的；迅速的
interfere with 干扰
melt down 熔化
explode [ɪk'spləʊd] *vt.* （使）爆炸

exploit ［ɪk'splɔɪt］ *vt.* 开采；开拓
profit ［'prɔfɪt］ *n.* ；*vi.* 有益；获利；收益；得益
protest ［'prəʊtest］ *vi.* ；*n.* 抗议；反对；申明
motivate ［'məʊtɪveɪt］ *vt.* 促动；激发；诱导；刺激
principle ［'prɪnsəpl］ *n.* 原则；原理

Oral practice（口语练习）

Q：Why are computer viruses called viruses?

A：Because they share some of the traits of biological viruses. A computer virus passes from computer to computer like a biological virus which passes from person to person.

Q：How do you protect yourself against viruses?

A：Running a more secure operating system like UNIX. Using virus protection software. Never double-click on an attachment that contains an executable that arrives as an e-mail attachment.

Q：What extention of the file is dangerous?

A：A file with an extension like EXE, COM or VBS is an executable, and an executable can do any sort of damage it wants. Once you run it, you have given it permission to do anything on your machine. The only defense is to never run executables that arrive via e-mail.

Exercises（练习题）

Ⅰ. Choices.

1. A computer virus is a _____ .
   A. living cell   B. factory   C. program   D. data file
2. _____ have no way to replicate automatically.
   A. E-mail viruses   B. Worms
   C. Shell viruses   D. Trojan horses
3. The only defense is to never run _____ that arrive via e-mail.
   A. Word files   B. spreadsheets
   C. executables   D. images (.GIF and .JPG)
4. The virus can only exist in a form that allows it to be executed as a _____ by the PC in some form.
   A. data   B. picture   C. e-mail   D. program
5. Which is not the way of the virus spreading?
   A. Over execute program.   B. Over a network.
   C. Over flash memory.   D. Over a mouse.

Ⅱ. Translate the following sentences into Chinese.

1. Do you run android antivirus software?

2. Rate of infection of virus of our country computer shows downtrend first.

3. Making sense of the data deluge.

4. The damage is becoming more apparent.

5. Hackers look for computers with security vulnerabilities and infect them in advance of an attack.

Ⅲ. Translate the following sentences into English.

1. 计算机病毒和黑客是有区别的。

2. 务必马上关掉那台计算机。

3. 病毒会损坏很多文件。

4. 有些病毒会导致计算机重新启动。

5. 我需要弄一些钱来支付杀毒软件的开销。

译文

计算机病毒

1. 计算机病毒定义

什么是计算机病毒？破坏计算机功能或者毁坏数据、影响计算机使用，并能自我复制的一组计算机指令或者程序代码。我们听说过的病毒有：电邮蠕虫病毒、网络钓鱼、灰鸽子。

计算机病毒是通过首次感染可执行文件或者系统硬盘和软盘系统区域，进行传播，并且能够自我复制的程序代码。病毒的种类有很多：木马病毒、蠕虫病毒、后门病毒、脚本病毒、宏病毒、系统病毒、捆绑机病毒、下载者病毒。以下是一些常见的病毒。

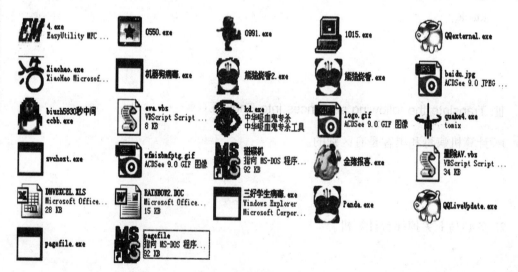

计算机病毒

2. 病毒是怎样传播的？

● 通过执行程序。

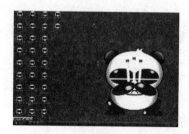

执行程序

Chapter 1 Basis of Computer

- 通过网络。

网络

- 通过可移动存储设备。

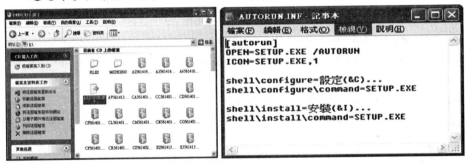

移动存储设备

病毒在整个互联网上传播的速度是极快的——仅需要一天!

3. 病毒在计算机上做什么?
- 破坏文件。
- 干扰电脑操作。
- 试图把自己传播得更广。

但要知道,病毒对硬件是不会有任何破坏的:不会熔化掉你的CPU,不会烧掉你的硬盘,也不会炸掉你的显示器。这里有一些方法可以防止病毒的侵害:重要的文件要备份;安装杀毒软件;定期扫描病毒。一些常用的杀毒软件有:360杀毒、瑞星、金山毒霸。

最后,介绍一下什么是黑客:在计算机安全方面,黑客是指专门寻找和利用计算机或计算机网络中的漏洞并加以攻击的人。使黑客活跃的原因多种多样:利益、抗议、挑衅。黑客这个词虽然也使用于一些与计算机安全无关的方面,比如特别懂计算机的人,或特别懂网络的人,但很少出现在主流语境中。他们或许非常擅长计算机,但是用错了方向,尽管我们仅仅是程序员,但是,我们也应该有原则。

Chapter 1

Basis of Computer

1-7 Artificial Intelligence (AI)

What is Artificial Intelligence (AI) exactly? As a beginning we offer the following definition:

AI is a branch of computer science concerned with the study and creation of computer systems that exhibit some form of intelligence: systems that learn new concepts and tasks, systems that can reason and draw useful conclusions about the world around us, systems that can understand a natural language or perceive and comprehend a visual scene, and systems that perform other types of feats that require human types of intelligence.

人工智能（AI）确切地说是什么？作为开始，我们提供下列定义：

AI 是计算机科学的一个分支，它涉及研究和创建显示某种形式智能的计算机系统：学习新概念和新任务的系统、能对我们周围的世界进行推理和得出有用结论的系统、能理解自然语言或理解和领会视觉场景的系统，以及执行需要人的各类智能的其他种类活动的系统。

Text（文本）

Artificial Intelligence

Dictionaries define intelligence as the ability to acquire, understand and apply knowledge, or the ability to exercise thought and reason. Of course, intelligence is more than this. It embodies all of the knowledge and feats, both conscious and unconscious, which we have acquired through study and experience: highly refined sight and sound perception; thought; imagination; the ability to converse, read, write, drive a car, memorize and recall facts, express and feel emotions; and much more.

Intelligence is the integrated sum of those feats which gives us the ability to remember a face not seen for thirty or more years, or to build and send rockets to the moon. It is those capabilities which set human beings apart from other forms of living things. And, as we shall see, the food for this intelligence is knowledge.

Can we ever expect to build systems which exhibit these characteristics? The answer to this

Chapter 1 Basis of Computer

question is yes! Systems have already been developed to perform many types of intelligent tasks, and expectations are high for near term development of even more impressive systems.

We now have systems which can learn from examples, from being told, from past related experiences, and through reasoning. We have systems which can solve complex problems in mathematics, in scheduling many diverse tasks, in finding optimal system configurations, in planning complex strategies for the military and for business, in diagnosing medical diseases and other complex systems, to name a few. We have systems which can "understand" large parts of natural languages. We have systems which can see well enough to "recognize" objects from photographs, video cameras and other sensors. We have systems which can reason with incomplete and uncertain facts. Clearly, with these developments, much has been accomplished since the advent of the digital computer.

In spite of these impressive achievements, we still have not been able to produce co-ordinated, autonomous systems which possess some of the basic abilities of a three-year-old child. These include the ability to recognize and remember numerous diverse objects from a scene, to learn new sounds and associate them with objects and concepts, and to adapt to many diverse new situations. These are the challenges now facing researchers in AI. And they are not easy ones. They will require important breakthroughs before we can expect to equal the performance of our three-year old.

In AI the goal is to develop working computer systems that are truly capable of performing tasks that require high levels of intelligence. The programs are not necessarily meant to imitate human senses and thought processes. Indeed, in performing some tasks differently, they may actually exceed human abilities. The important point is that the systems are all capable of performing intelligent tasks effectively and efficiently.

Key words and expressions（重点词汇）

Artificial Intelligence 人工智能
a branch of 分支
exhibit [ɪgˈzɪbɪt] vt. 呈现；陈列；展览；证明
conclusion [kənˈkluːʒn] n. 结论
perceive [pəˈsiːv] v. 理解；意识到；察觉
comprehend [ˌkɒmprɪˈhend] vt. 理解；领会；包含
diverse [daɪˈvɜːs] adj. 不同的；多种多样的
strategy [ˈstrætədʒi] n. 策略；战略
diagnose [ˈdaɪəgnəʊz] vt. 诊断；判断
advent [ˈædvent] n. 出现
exceed [ɪkˈsiːd] vt. 超过；超越；胜过

Expanded vocabulary（扩展词汇）

feat [fiːt] n. 技术，本领；功绩
embody [ɪmˈbɒdi] vt. 表现

conscious ['kɔnʃəs] *adj.* 有意识的
refine [rɪ'faɪn] *vt.* 提炼；改善
schedule ['ʃedjuːl] *vt.* 安排
autonomous [ɔː'tɔnəməs] *adj.* 自治的；有自主权的
in spite of 虽然；尽管

Oral practice（口语练习）

1. Like other definitions of complex topics, an understanding of AI requires an understanding of related terms such as intelligence, knowledge, reasoning, thought, cognition, learning, and a number of computer-related terms. While we lack precise scientific definitions for many of these terms, we can give general definitions of them. And, of course, one of the objectives of this text is to impart special meaning to all of the terms related to AI, including their operational meanings.

2. To gain a better understanding of AI, it is also useful to know what AI is not. AI is not the study and creation of conventional computer systems. Even though one can argue that all programs exhibit some degree of intelligence, an AI program will go beyond this in demonstrating a high level of intelligence to a degree that equals or exceeds the intelligence required of a human in performing some tasks.

3. AI is not the study of the mind, nor of the body, nor of languages, as customarily found in the fields of psychology, physiology, cognitive science, or linguistics. To be sure, there are some overlaps between these fields and AI. All seek a better understanding of the human's intelligence and processes.

Exercises（练习题）

Ⅰ. Questions and answers.

1. What is AI?

2. Please list some examples of AI in our life.

3. Why is AI not necessarily meant to imitate human senses and thought processes?

4. What is the goal of AI?

5. What kind of AI do you want?

Ⅱ. Translate the following paragraph into Chinese.

Finally, a better understanding of AI is gained by looking at the component areas of study that make up the whole. These include such topics as robotics, memory organization, knowledge representation, storage and recall, learning models, inference techniques, commonsense reasoning, dealing with uncertainty in reasoning and decision making, understanding natural language, pattern recognition and machine vision methods, search and matching, speech recognition and synthesis, and a variety of AI tools.

Ⅲ. Translate the following paragraph into English.

AI 正进入这样的年代,即实用的商业产品现在已可用了,包括各种各样的机器人设备、识别形状和对象的视觉系统、执行很多困难任务的专家系统、帮助调整学生的学习,并监控学生学习进度的智能教育系统、帮助用户建造专门知识库的"智能"编辑器,以及能学习以改进其性能的一些系统。

人工智能

字典把智能定义为获得、理解和应用知识的能力,或者是实行思维和推理的能力。当然,智能不只是这点。它具体体现了有意识地和无意识地通过学习和经验获得的所有知识和技艺:高度精确的视觉和听觉感知;思维;想象;交谈、读、写、驾车、记忆和回忆事实、表达和感受情感的能力,以及更多。

智能是这些技艺的集成之和,使我们能回忆起 30 年或更多年未见的面孔,或建造并发送火箭到月球的能力。是这些能力使人类区别于其他的有生命体。并且,如同我们将看到的,这个智能的食粮是知识。

我们会一直期望建造显示这些特征的系统吗?此问题的答案是 yes!一些系统早已被开发来执行很多种类的智能任务,并且对近期开发给人印象更为深刻的系统寄予了很高的期望。

现在有一些系统能从例子、从被告知的、从过去相关的经验和通过推理进行学习。有一些系统能解决数学方面的、调度多种多样任务方面的、寻找最佳系统配置方面的、计划军事和商业的复杂策略方面的、诊断医学疾病方面的复杂问题,还有其他一些复杂系统,这里仅举几个例子。有些系统能"理解"一些自然语言的大部分,有些系统的视觉好得足以"识别"照片上、摄像机和其他传感器拍摄的图像上的物体。有些系统能够以不完备的和不确定的事实进行推理。显然，关于这些开发，自数字计算机问世以来，很多已完成了。

  尽管有这些印象深刻的成就，我们仍然不能生产具有三岁小孩有的某些基本能力的协调而自主的系统，这些包括识别和记忆一个景象中众多的各种各样的对象、学习新的声音和把它们同对象与概念相关联，以及欣然适应多种多样的新情况的能力。这些都是 AI 研究人员现在面临的挑战，并且它们都不是容易解决的问题。在我们能期望比得上三岁小孩的性能之前，它们将需要一些重要的突破。

  就 AI 而论，目标是开发一些运转的计算机系统，它们真正能执行一些需要高级智能的任务。程序未必打算模仿人的感觉和思维过程。确实，在以不同的方式执行某些任务的过程中它们会实际上超过人的一些能力。重要的一点是这些系统都能有效而高效率地执行智能任务。

Chapter 2

Network

2-1 Local Area Network

Now people's lives become increasingly dependent on the network. We can read the news and know the world through the network. There are all kinds of network in our life, such as transportation network, telecommunication network, television network, electricity network, and postal network, etc.

现在人们的生活越来越离不开网络,通过网络我们可以读新闻、知天下。我们的生活中有各种各样的网络,如交通网、通信网、电视网、电力网、邮政网等。

Local Area Network

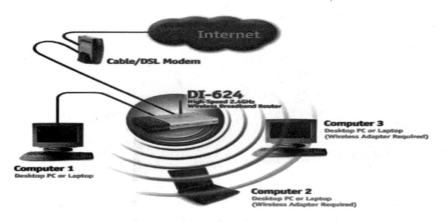

Local area network

Local Area Network (LAN) is a computer network covering only a small physical area, like a home, an office, or a small group of buildings. Its geographical scope is between 10m and 1km and the transmission rate is usually from 1Mbps to 20Mbps. The defining characteristics of LANs, in contrast to WANs, usually include higher data-transfer rates, smaller distance in geographic areas and less cost in telecommunication lines.

LAN consists mainly of network hardware and software. Hardware systems include computers (servers, mini computers, terminals), Inter Connect Equipments (ICEs), and transmitting media.

The computer is the main equipment of network, and different computers bear different tasks in network. The server is designed to provide services to clients. It affects the overall performance of the network, while a terminal is an electromechanical hardware device that is used for entering data into, and displaying data from a computer or a computing system. You can manually configure the IP address for the PC, and you can also automatically assign an appropriate address via Dynamic Host Configuration Protocol (DHCP) for each host.

ICES should be used to connect different computers or network-related accessories, such as network interface cards, integrated devices, switches, repeaters, routers, modems.

Network interface card

Integrated device

Switch

Router

Modem

Chapter 2 Network

There are many different types of LANs, Ethernets being the most common for PCs. Each LAN has its own unique topology (the geometric arrangement of devices on a LAN), or geometric arrangement. Pictures below show three basic topologies: bus, ring, and star:

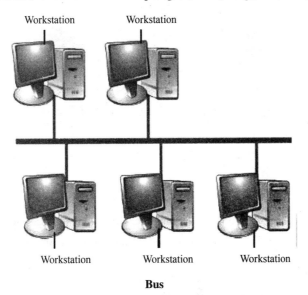

Bus

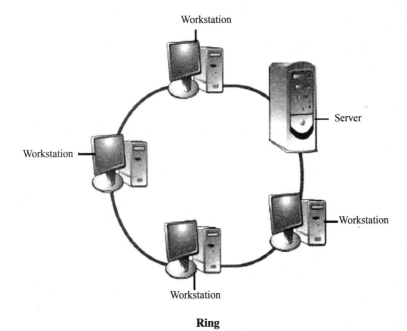

Ring

In bus network topology, each computer or server is connected to the single bus cable through some kind of connector. In local area networks where the star topology is used, each machine is connected to a central hub. In contrast to the bus topology, the star topology allows each machine on the network to have a point to point connection to the central hub. In ring network topology, each computer is connected to the network in a closed loop or ring. Each machine or computer has a

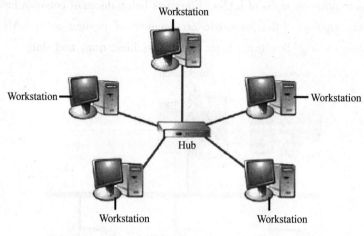

Star

unique address that is used for identification prupose. The signal passes through each machine or computer connected to the ring in one direction. Token ring is the most famous ring network topology.

Key words and expressions（重点词汇）

mini computer 小型机
terminal ['tə:mɪnl] n. 终端；终点站
inter connection equipment 互联设备
transmitting media 传输设备
client ['klaɪənt] n. 客户端；顾客；客户
assign [ə'saɪn] vt. 分派，指定；指派（过去式 assigned，过去分词 assigned，
    现在分词 assigning，第三人称单数 assigns）
accessories [ək'sesərɪz] n. （accessory 的名词复数）附属设备；配件
topology [tə'pɔlədʒɪ] n. 拓扑结构；拓扑
Ethernet ['i:θənet] n. 以太网；以太网络
bus [bʌs] 总线型
start [stɑ:t] 星型
ring [rɪŋ] 环型
cable ['keɪbl] 电缆
hub [hʌb] 集线器

Expanded vocabulary（扩展词汇）

geographical [ˌdʒi:ə'græfɪkl] adj. 地理学的；地理的
scope [skəup] n. 范围；适用范围；作用域
physical ['fɪzɪkl] adj. 物理的；物质的
transmission [træns'mɪʃn] n. 播送；传送；传输

in contrast to 与……相反；与……形成对照；与……成对比
connecter ［kə'nektə］ *n.* 连接者；连接物
telecommunication ［ˌtelɪkəˌmjuːnɪ'keɪʃn］ *n.* 电信；电通信
performance ［pə'fɔːməns］ *n.* 性能；表演；演出；绩效
electromechanical ［ɪ'lektrəʊmɪ'kænɪkəl］ *adj.* 电动机械的；机电的；电机的
appropriate ［ə'prəʊpriət］ *adj.* 适当的；恰当的；合适的 *v.* 盗用；侵吞（过去式 appropriated，过去分词 appropriated，现在分词 appropriating，第三人称单数 appropriates）
geometric ［ˌdʒiːə'metrɪk］ *adj.* 几何学的；几何体的
arrangement ［ə'reɪndʒmənt］ *n.* 安排；排列；布置
identification ［aɪˌdentɪfɪ'keɪʃn］ *n.* 识别；认同；鉴别；标识

Oral practice（口语练习）

Q：What are you doing now?
A：I am on Internet.
Q：Is that interesting?
A：Surely, you can find a lot of things on it.
Q：What is Internet used for?
A：I can use it to make friends, watch movies or do shoppings.
Q：How amazing it is!
A：Yes, you can try it yourself.

Exercises（练习题）

Ⅰ. Choices.

1. _____ is used to communicate with another computer over telephone lines.
   A. Keyboard   B. Modem   C. Mouse   D. Printer
2. Alice received an invitation from her boss, _____ came as a surprise.
   A. it   B. that   C. which   D. he
3. The weather turned out to be very good, _____ was more than we could expect.
   A. what   B. which
   C. that   D. it
4. I don't like _____ as you read.
   A. the novels   B. the such novels
   C. such novels   D. same novels
5. This is one of the best films _____ .
   A. that have been shown this year   B. that have shown
   C. that has been shown this year   D. that you talked

Ⅱ. Translate the following sentences into Chinese.

1. But now, computer has been widely used in almost every field.

2. TCP/IP is a communication protocol, which provides many different networking services.

3. The boy, whose father is an engineer, studies very hard.

4. These apple trees, which I planted three years ago, have not borne any fruit.

5. They went to London, where they lived for six months.

Ⅲ. Translate the following sentences into English.

1. 广域网是国家间的网络。

2. 计算机网络传送数据。

3. 他们很感激 Tom，没有他的支持他们是不会成功的。

4. 众所周知，地球是圆的。

5. 她总是马虎大意，我们可别这样。

Chapter 2　Network

局　域　网

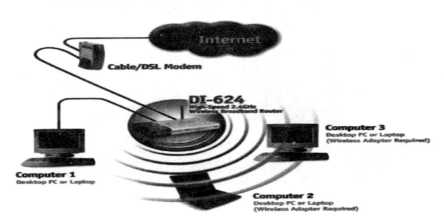

局域网拓扑图

局域网的地理范围小，分布在诸如一个家庭、一个办公室或小建筑群。通常，它的地域范围在 10～1 000 m，传输率在 1～20Mb/s。和广域网相比较，局域网的特点是传输速度快、分布距离近、连接费用低。

局域网主要由网络硬件和网络软件组成。网络硬件系统包括计算机（服务器、小型机、终端）、互联设备和传输介质等。

计算机是构成网络的主要设备，不同的计算机在网络中担负着不同的工作任务。服务器的作用是为客户提供服务，它影响网络的整体性能；而终端则是一台电子计算机或者计算机系统，用来让用户输入数据并显示计算结果的机器。可以通过手动方式为 PC 机配置 IP 地址，也可以通过动态主机配置协议（DHCP）为各主机自动分配一个合适的地址。

互联设备是用于连接网络中不同计算机或者网路的相关配件，主要包括网络接口卡、集线器、交换机、中继器、路由器、调制解调器等。

网卡

集成器

交换机

路由器

调制解调解

有多种不同类型的局域网。以太网是PC机最常使用的一种。每一个局域网都有它自己的拓扑结构或几何分布。下面是三种基本的拓扑结构：总线型、环形以及星型。

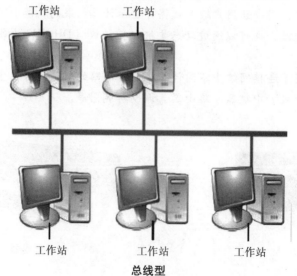

总线型

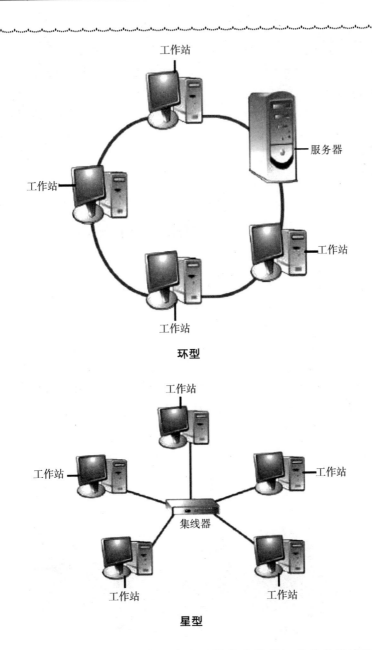

　　在总线拓扑结构中，可以通过某种连接器将每台计算机或服务器连接到单总线电缆。通常，星型拓扑结构用在局域网中。在这种结构中，网络中的每台机器通过中央集线器进行连接。在环型拓扑结构中，每台计算机连接到闭合环路。每一台机器或计算机都有一个唯一的识别地址。信号通过每台计算机按单一方向流动。令牌环是最著名的环形拓扑结构。

Chapter 2
Network

2 – 2　Internet

Introduction（导读）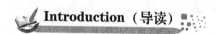

Through the Internet, we can send and receive mail and QQ chat. Through the Internet, we can watch movies, listen to music and enjoy entertainment. Through the Internet, we can also buy things, create and share micro-blogs....

通过网络，我们可以收发邮件、QQ 聊天；通过网络，我们可以看电影、听音乐、享受娱乐；通过网络，我们还可以买东西、搞创作、分享微博……

Text（文本）

Internet

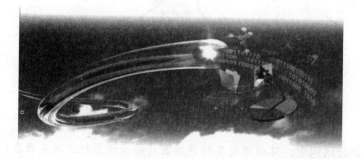

Internet

Internet is a giant network of computers located all over the world that communicate with each other. Internet is an international collection of computer networks that all understand a standard system of addresses and commands, connected together through backbone systems. It was started in 1969, when the U. S. Department of Defence established a nationwide network to connect a handful of universities and contractors. The original idea was to increase computing capacity that could be shared by users in many locations and to find out what it would take for computer networks to survive a nuclear war or other disaster by providing multiple paths between users.

Over the years, additional networks joined which added access to more and more computers. The first international connections, to Norway and England, were added in 1973. Today thousands of networks and millions of computers are connected to Internet. It is growing so quickly that nobody can say exactly how many users "On the Net."

Internet is the largest repository of information which can provide very very large network resources. The network resources can be divided into network facilities resources and network information resources. The network facilities resources provide us with the ability of remote computation and communication. The network information resources provide us with all kinds of information services, such as science, education, business, history, law, art, and entertainment, etc.

The most commonly used network service is electronic mail (E-mail), or simply as mail. Mail permits network users to send textual messages to each other. Computers and networks handle delivering the mail, so that communicating with mail users does not have to handle details of delivery, and does not have to be present at the same time or place. The simplest way to access a file on another host is to copy it across the network to your local host. FTP can do this.

World Wide Web (WWW) is a networked hypertext protocol and user interface. It provides access to multiple services and documents. A jump to other Internet service can be triggered by a mouse click on a "hot linked" word, image, or icon on the Web page.

As more and more systems join Internet, and as more and more forms of information can be converted to digital form, the amount of stuff available to Internet users continues to grow, so we can say that Internet is your PC's window to the rest of the world.

Key words and expressions（重点词汇）

giant ['dʒaɪənt] n. 巨人；大汉；adj. 特大的；巨大的
all over the world 全世界；世界各地
collection [kə'lekʃn] n. 收集；采集；集合；收集物
standard ['stændəd] n. 标准；规格　adj. 标准的；合格的
backbone ['bækbəʊn] n. 主干；支柱
handful ['hændfʊl] n. 少数；少量
contractor [kən'træktə(r)] n. 订约人；承包人；协议者
original [ə'rɪdʒənl] adj. 原始的；独创的；最初的　n. 原文；原型
share [ʃeə(r)] vi. 分享；分担
path [pɑːθ] n. 小路；路；路线；路程
access ['ækses] vt. 接近；进入
facility [fə'sɪlɪti] n. 设备；容易；条件
remote [rɪ'məʊt] n. 远程操作；遥控器　adj. （时间上）遥远的；远离的；远程的
handle ['hændl] vi. 操作；操控（复数 handles，过去式 handled，过去分词 handled，
　　　　　现在分词 handling，第三人称单数 handles）

deliver [dɪˈlɪvə(r)] vt. 发表；递送　vi. 投递；传送（过去式 delivered，过去分词 delivered，现在分词 delivering，第三人称单数 delivers）

present [ˈpreznt] adj. 现在的；目前的　vt. 介绍；出现；提出；赠送

document [ˈdɔkjumənt] n. （计算机）文档；证件；公文　vt. 证明；记录

icon [ˈaɪkɔn] n. 图标；图示

digital form 数字形式；数码形式

Expanded vocabulary（扩展词汇）

communicate [kəˈmjuːnɪkeɪt] vt. 传达；表达（过去式 communicated，过去分词 communicated，现在分词 communicating，第三人称单数 communicates）

international [ˌɪntəˈnæʃnəl] adj. 国际的；国际关系的（复数 internationals）

establish [ɪˈstæblɪʃ] vt. 建立；创建（过去式 established，过去分词 established，现在分词 establishing，第三人称单数 establishes）

locate [ləʊˈkeɪt] vt. 位于；确定……的位置　vi. 定位；（过去式 located，过去分词 located，现在分词 locating，第三人称单数 locates）

universities [ˌjuːnɪˈvɜː(r)sətiz] n. 大学（university 的名词复数）

survive [səˈvaɪv] vi. 幸存；活下来（过去式 survived，过去分词 survived，现在分词 surviving，第三人称单数 survives）

multiple [ˈmʌltɪpl] adj. 多重的；多个的　n. <数>倍数；多个

repository [rɪˈpɔzətri] n. 仓库；贮藏室；资源库；储存库

additional [əˈdɪʃənl] adj. 额外的；附加的；另外的；追加的

computation [ˌkɔmpjuˈteɪʃn] n. 计算；估计

multiple [ˈmʌltɪpl] adj. 多重的；多个的　n. 多个；倍数

stuff [stʌf] n. 材料；原料；资料

hot link 热链接

trigger [ˈtrɪgə(r)] vt. 引发，触发（复数 triggers，过去式 triggered，过去分词 triggered，现在分词 triggering，第三人称单数 triggers）

Oral practice（口语练习）

Q: Hello, Tom! What are you doing?

A: Quick test, and review here.

Q: Then, you progress now?

A: Little progress, an English test is approaching. Really do not know how to do! How is your English and tell me how you learn English?

Q: In fact, you can get on the Internet, online listening to English songs, seeing the film in English. This can increase your interest in learning English. It will be helpful for your English learning.

A: Really? Is it really helpful? I didn't know that before.

Q: It is, of course. In online classroom you can learn English, cognize Western cultures and expand knowledge.

A: That is really wonderful. The English language learning could be with so much fun.

Q: Right. In fact, learning English is not so difficult as I think.

A: Next time, please teach me how to learn English on the Internet.

Q: Yes, I am now going to school, and I wish you success in examinations. Byebye!

A: Well, see you next time! Byebye!

Exercises (练习题)

I. Choices.

1. A router reads _____ address on a packet to determine the next hop.
   A. IP   B. MAC
   C. source   D. ARP

2. The network resources can be divided into network _____ and network information resources.
   A. facilities resources   B. finite resources
   C. diminishing resources   D. limited resources

3. I enjoyed the movie very much. I wish I _____ the book from which it was made.
   A. have read   B. had read
   C. should have read D. are reading

4. The two students talked as if they _____ friends for years.
   A. should be   B. would be
   C. have been   D. had been

5. It is important that I _____ with Mr. Williams immediately.
   A. speak   B. spoke
   C. will speak   D. to speak

II. Translate the following sentences into Chinese.

1. The goal of your use of the Internet is exchanging messages or obtaining information.

2. The Internet is a huge interconnected system, but it uses just a handful of methods to move data around.

3. If somebody else had something interesting stored on their computer, it was a simple matter to obtain a copy (assuming the owner did not protect it). The damage is becoming more apparent.

4. Life on-line can be a much richer experience when you aren't restricted to just written words and still pictures.

5. We have already known that an Internet computer is identified by its IP address, which is difficult to remember.

Ⅲ. Translate the following sentences into English.

1. BBS 是一个电子信息中心。

2. 最新的一种发送信息的方式就是电子邮件。

3. 通常来说，邮件到达目的地只需要几秒钟。

4. 万维网主要使用 HTTP（超文本传输协议）。

5. HTML 是创建万维网上文档的混合用语。

译 文

互 联 网

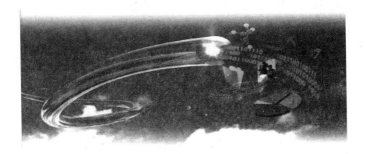

互联网

  Internet 是由位于世界各地相互通信的计算机连接而成的巨大的计算机网络。Internet 是计算机网络的国际性的集合，这些网络都符合具有地址和命令的标准体系，并经骨干网连在一起。Internet 始建于 1969 年，当时美国国防部为连接少数几所大学和协议企业而建立了一个全国性网络。最初的想法是要增加计算机能力并可由许多地点的用户共享，并且通过提供用户间多条路径来找到哪一种计算机网络能够在核战或其他灾难中幸存。

  几年间，新网络的接入使越来越多的计算机加入进来。在 1973 年进行了第一次与挪威和英国的国际连接。今天，有成千上万的计算机网络和数百万台计算机与 Internet 相连。Internet 发展如此之快以至于没有人能准确地说出网上有多少用户。

  Internet 是最大的信息宝库，它可以提供非常巨大的网络资源。这种网络资源可分为网络设备资源和网络信息资源。网络设备资源使我们能够进行远程计算和通信。网络信息资源向我们提供各种各样的信息服务，如科学、教育、商务、历史、法律、艺术和娱乐等。

  最常使用的网络服务是电子邮件，或简称邮件。电子邮件允许网络用户彼此传送文本消息。邮件的传递由计算机和网络处理，邮件用户不必关心传递的细节，也不必同时在场。从其他主机中获得文件的最简单的方式是通过网络将其复制到你的计算机上。文件传送（FTP）可完成这项工作。

  全球网（WWW）是一种网络的超文本协议和用户界面。它提供多种服务和文件接入方法。向 Internet 其他服务的跳转可在"网"页上用鼠标器点击"热链接"的字、图像或按钮来启动。

  随着越来越多的系统加入 Internet，同时随着越来越多的信息可以转变成数字形式，Internet 用户所能得到的东西也在继续增加。所以我们可以说，Internet 是你的 PC 机通向世界其他地方的窗口。

Chapter 2

Network

2-3 Mobile Communication

 **Introduction**（导读）

  With the development of science and technology, communication technology has been improved rapidly and the heterogeneity of mobile communication system has been seen. The trend of future mobile communication is to realize the combination of the mobile communication networks, in order to serve the users better.

  随着科技的进步，通信技术得到了飞速的发展，移动通信系统的异构性更加充分。未来移动通信发展的趋势是实现移动通信网络的融合，更好地为用户服务。

 **Text**（文本）

Mobile Communication

  There is one or two people moving in the communication, which is called mobile communication. It includes land, sea and air mobile communications. Recalling the development of mobile communications, we can know it has experienced four stages.

  The first generation of mobile communication technology is called 1G. Only the voice traffic can be transmitted by the technology and it has the network capacity constraint.

  The second generation mobile communication technology is called 2G with digital voice transmission technology as the core. Generally it can not be used to transmit the information just as e-mail and software directly. Instead we can call each other and transmit the information such as time and date.

CPU

  3G is the third generation mobile communication technology. The information both voice and data can be transmitted by 3G. 3G is a new combination of wireless communication systems with wireless and Internet multimedia communication.

3G

4G is the fourth generation mobile communication technology. At present this technology has not been accepted by the next generation wireless communication standard actually. Therefore we can just call it 3.9G. 4G is the combination of 3G and WLAN and it can be used to transmit the data, audio, video and image rapidly.

In general, we should do something based on 3G and realize the transition from 3G to 4G, in order to reach the standard of the real 4G.

4G

Key words and expressions（重点词汇）

  user ［'juːzə(r)］ *n.* 用户；用户的；使用者；户名
  stage ［steɪdʒ］ *n.* 阶段；舞台
  combination ［ˌkɒmbɪ'neɪʃn］ *n.* 组合；结合
  information ［ˌɪnfə'meɪʃn］ *n.* 信息
  voice ［vɔɪs］ *n.* 声音；语音
  generation ［ˌdʒenə'reɪʃn］ *n.* 一代；代
  transmit ［træns'mɪt］ *vt.* 传送；传播；传输
  wireless ［'waɪələs］ *adj.* 无线的
  WLAN ［'dʌbljulæn］ *n.* 无线局域网
  transition ［træn'zɪʃn］ *n.* 过渡；转变
  traffic ［'træfɪk］ *n.* 流量；交通
  standard ［'stændəd］ *n.* 标准

Expanded vocabulary（扩展词汇）

  heterogeneous ［ˌhetərə'dʒiːniəs］ *adj.* 各种各样的；成分混杂的；异构的
  recall ［rɪ'kɔːl］ *vt.* 召回；使想起，回想；取消；调回工厂
  experience ［ɪk'spɪəriəns］ *vt.* 感受；亲身参与，亲身经历；发现
  capacity ［kə'pæsəti］ *n.* 容量；性能；才能；生产能力；*adj.* 充其量的，最大限度的
  constraint ［kən'streɪnt］ *n.* 约束；限制；强制
  digital ［'dɪdʒɪtl］ *adj.* 数字的；数据的
  core ［kɔː(r)］ *n.* 中心，核心，精髓

Oral practice（口语练习）

Bill： Hi, Ellen! Could you tell me what mobile communication is?
Ellen： Well, there is one or two people moving in the communication, which is called mobile communication.
Bill： Oh, can you tell me something about its experience?
Ellen： Of course! It has experienced four stages: 1G, 2G, 3G and 4G.
Bill： I have heard of it. The first generation of mobile communication technology is called 1G. The second is called 2G, and the third, the fourth...

Ellen: Yeah! 4G is the latest mobile communication at present!
Bill: 4G is the combination of 3G and WLAN and it can be used to transmit the data, audio, video and image rapidly.
Ellen: We can talk a lot about the mobile communication in future when we are free!
Bill: It's very good!

Exercises (练习题)

Ⅰ. Choices.

1. The mobile communications has experienced _____ stages.
   A. one   B. two   C. three   D. four
2. _____ is the newest technology.
   A. 2G   B. 3G   C. 4G   D. 2.5G
3. The information both voice and _____ can be transmitted by 3G.
   A. data   B. e-mail   C. image   D. video
4. _____ is the short name of the first generation mobile communication.
   A. 1G   B. 2G   C. 3G   D. 4G
5. The information both voice and _____ can be transmitted by 3G.
   A. data   B. e-mail   C. image   D. video

Ⅱ. Translate the following sentences into Chinese.

1. The mobile communication is used widely.

2. Could you give me an example for the mobile communication?

3. We have made great progress in the technology.

4. We must make great progress on the way to 4G.

5. I have a 4G SIM card.

Chapter 2　Network

Ⅲ. Translate the following sentences into English.

1. 你看了关于4G的新广告了吗?

2. 4G是个新生事物，发展还不够成熟。

3. 我们对3G都非常熟悉。

4. 我们都应该为移动通信的发展做出努力。

5. 移动通信的出现改变了我们的生活。

移 动 通 信

　　通信双方有一方或两方处于运动中的通信称为移动通信。它包括陆、海、空移动通信。回顾移动通信的发展历程，移动通信的发展大致经历了四个阶段。
　　1G表示第一代移动通信技术。1G无线系统在设计上只能传输语音流量，并受到网络容量的限制。
　　2G表示第二代手机通信技术，以数字语音传输技术为核心。一般定义为无法直接传送如电子邮件、软件等信息；只可以用来通话和传送一些如时间、日期等的信息。
　　3G是第三代移动通信技术。3G服务能够同时传送声音及数据信息。3G是指将无线通信与国际互联网等多媒体通信结合的新一代移动通信系统。
　　4G指的是第四代移动通信技术。目前该技术其实并未被下一代无线通信标准所认可，因此我们只能称其为3.9G。4G是集3G与WLAN于一体，并能够快速传输数据、音频、视频和图像等的一种通信技术。

2G

3G　　　　　　　　　　　　　　4G

总的来说，要全面顺利地实现4G通信，需直接在3G通信网络的基础设施之上，采用逐步引入的方法，使移动通信从3G逐步向4G过渡。

Chapter 2

Network

2-4 Internet of Things

Introduction（导读）

The English name of Internet of Things, is referred to as IOT. Internet of Things works through the sensor, radio frequency identification technology, global positioning system technology, makes real-time acquisition of any monitoring, connectivity, interactive objects or processes, collects their sound, light, heat, electricity, mechanics, chemistry, biology, the location of a variety of the information you need to realize through a variety of possible things and things, objects and people in the Pan-link intelligent perception of items and processes, identification and management.

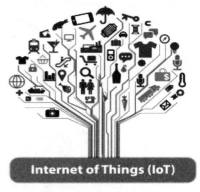

Internet of Things

物联网的英文名称为 Internet of Things，简称 IOT。物联网通过传感器、射频识别技术、全球定位系统等技术，实时获得任何需要监控、连接、互动的物体或过程方面的信息，采集其声、光、热、电、力学、化学、生物、位置等你需要的信息，通过各类可能的网络接入，实现物与物、物与人的泛在链接，实现对物品和过程的智能化感知、识别和管理。

Text（文本）

Internet of Things

Internet of Things is constructed on the basis of the Internet, RFID, wireless data communications technology, to cover everything in the world. In this network, the goods (products) can "exchange" with each other, without the need for human intervention. Its essence is the use of radio frequency identification (RFID) technology to achieve the interconnection and sharing of the automatic identification of goods (products) and information through the computer Internet.

The Internet of Things, a very important technology, is radio frequency identification (RFID) technology. RFID, the abbreviation of radio frequency identification (Radio Frequency

Identification) technology, is an automatic identification technology emerging in the 1990s and the more advanced non-contact identification technology. The development of RFID technology is based on a simple RFID system, combined with existing network technology, database technology, middleware technology, to build IOT composed by a large number of networked readers and numerous mobile labels, much larger than the Internet.

RFID, is a technique able to let items "speak." In the "Internet of Things" concept, RFID tags store the specification and interoperability information collected automatically by wireless data communications network to a central information system, to achieve the identification of goods (products), and then through the open computer network to realize information exchange and sharing, and "transparent" management.

The information technology revolution in the Internet of Things is referred to as IT mobile Pan of a specific application. With Internet of Things through Intellisense, identification technology and pervasive computing, ubiquitous network convergence applications, breaking the conventional thinking before, human beings can achieve ubiquitous computing and network connectivity. The traditional thinking has been the separation of physical infrastructure and IT infrastructure: on the one hand, airports, roads, buildings, while on the other hand, the data center, PC, broadband. In the era of the "Internet of Things," reinforced concrete, cable with the chip, broadband integration into a unified infrastructure, in this sense, the infrastructure is more like a new site of the Earth, and the world really works on it, which includs economic management, production operation, social and even personal life. Internet of Things can be found in the personal health, smart grid, public transportation and other aspects of the extremely wide range of applications. As long as an object is embedded in a specific radio frequency tags, sensors and other devices connected to the Internet,

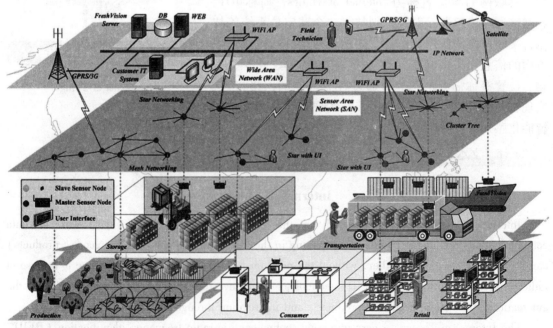

The application of Internet of Things

Chapter 2　Network

will be able to form a large network systems. In this line, even thousands of miles away, people can easily learn and control the object information.

"Internet of Things" makes it much more refined and dynamic to manage production and life, to manage the future of the city, to achieve the status of "wisdom", to improve resource utilization and productivity levels, and to improve the relationship between man and nature. Some experts predict that within 10 years things may be a large-scale popularity and become a trillion-scale high-tech market. Then, in the personal health, traffic control, environmental protection, public safety, peace at home, smart fire, industrial monitoring, elderly care, and almost all the other areas, "Internet of Things" will play a role. Some experts said that only three to five years' time, things would make a full access to people's lives, and change people's way of life.

Key words and expressions（重点词汇）

sensor ['sensə] *n.* 传感器
acquisition [ˌækwɪ'zɪʃ(ə)n] *n.* 获得物；获得
monitoring ['mɔnɪtərɪŋ] *n.* 监视
intervention [ˌɪntə'venʃ(ə)n] *n.* 介入
abbreviation [əˌbriːvɪ'eɪʃ(ə)n] *n.* 缩写
specification [ˌspesɪfɪ'keɪʃ(ə)n] *n.* 规格
Intellisense [in'telisens] *n.* 智能感知
ubiquitous [juː'bɪkwɪtəs] *adj.* 普遍存在的；无所不在的
conventional [kən'venʃənl] *adj.* 传统的；常见的
grid [grɪd] *n.* 输电网
embedd [ɪm'bed] *v.* 嵌入

Expanded vocabulary（扩展词汇）

Pan-link　*n.* 泛在链接
essence ['es(ə)ns] *n.* 本质；实质
middleware ['mɪdlwɛə] *n.* 中间件；中间设备
infrastructure ['ɪnfrəstrʌktʃə] *n.* 基础设施
reinforced [ˌriːɪn'fɔːst] *adj.* 加固的；加强的
concrete ['kɔŋkriːt] *n.* 具体物；凝结物
refined [rɪ'faɪnd] *adj.* 精炼的；精确的；微妙的
utilization [ˌjuːtɪlaɪ'zeɪʃən] *n.* 利用；使用

Oral practice（口语练习）

Q：What does IOT mean?
A：It is the abbreviation of Internet of Things.
Q：How does it work?
A：It works through the sensor, radio frequency identification technology, global positioning system

to collect and process information.

Q: What is it used for?

A: In the personal health, traffic control, environmental protection, public safety, peace at home, smart fire, industrial monitoring, elderly care, and almost all the other areas, "Internet of Things" will play a role.

Exercises (练习题)

Ⅰ. Short-answer questions.

1. What does this text talk about?

2. Please list the techniques in our life IOT refers to.

3. How does RFID work?

4. What will IOT be come in the future?

Ⅱ. Translate the following paragraphs into Chinese.

1. According to the literal meaning of explanation of things, Internet of Things, also known as the sensor network, refers to a variety of information sensing devices and the Internet combined to form a huge network, will enable all of the items and network connections to facilitate the identification and management.

2. As a high-tech market, the IOT will form three major market segments in the public management and service, enterprise, individual and family application in the near future.

Ⅲ. Translate the following sentences into English.

1. 每一次危机，都会催生一些新技术，而新技术也是使经济，特别是工业走出危机的巨大推动力。

2. 席卷全球的金融危机也在催生新的经济驱动力诞生，物联网就是众人最为推崇的动力。

译文

物　联　网

物联网是在计算机互联网的基础上，利用RFID、无线数据通信等技术，构造一个覆盖世界上万事万物的"Internet of Things"。在这个网络中，物品（商品）能够彼此进行"交流"，而无须人的干预。其实质是利用射频自动识别（RFID）技术，通过计算机互联网实现物品（商品）的自动识别和信息的互联与共享。

物联网中非常重要的技术是射频识别（RFID）技术。RFID是射频识别（Radio Frequency Identification）技术的英文缩写，是20世纪90年代开始兴起的一种自动识别技术，是目前比较先进的一种非接触识别技术。其发展是以简单RFID系统为基础，结合已有的网络技术、数据库技术、中间设备技术等，构筑一个由大量联网的阅读器和无数移动的标签组成的，比Internet更为庞大的物联网。

而RFID，正是能够让物品"开口说话"的一种技术。在"物联网"的构想中，RFID标签中存储着规范而具有互用性的信息，通过无线数据通信网络把它们自动采集到中央信息系统，实现物品（商品）的识别，进而通过开放性的计算机网络实现信息交换和共享，实现对物品的"透明"管理。

物联网信息化革命的浪潮被称为信息技术移动泛在化的一个具体应用。物联网通过智能感知、识别技术与普适计算、泛在网络的融合应用，打破了之前的传统思维，人类可以实现无所不在的计算和网络连接。传统的思路一直是将物理设施和IT设施分开：一方面是机场、公路、建筑物，而另一方面是数据中心、个人电脑、宽带等。而在"物联网"时代，钢筋混凝土、电缆将与芯片、宽带整合为统一的设施，在此意义上，基础设施更像是一块新的地球工地，世界的运转就在它上面进行，其中包括经济管理、生产运行、社会管理乃至个人生活。"物联网"在个人健康、智能电网、公共交通等方面的应用范围极其广泛。只要将特定物体嵌入射频标签、传感器等设备，与互联网相连后，就能形成一个庞大的联网系统，在这个网上，即使远在千里之外，人们也能轻松获知和掌控物体的信息。

"物联网"使得人们可以以更加精细和动态的方式管理生产和生活，管理未来的城市，达到"智慧"状态，提高资源利用率和生产力水平，改善人与自然间的关系。有专家预测10年内物联网就可能大规模普及，发展成为上万亿规模的高科技市场。届时，在个人健康、交通控制、环境保护、公共安全、平安家居、智能消防、工业监测、老人护理等几乎所有领域，物联网都将发挥作用。有专家表示，只需3~5年时间，物联网就会全面进入人们的生活，改变人们的生活方式。

Chapter 2
Network

2-5 Protocols

Introduction (导读)

At the classification of networks for data transmission to be successful, the sender and the receiver must follow a set of communication rules for the exchange of information. These rules for exchanging data between computers are known as the protocol. The rules can be understood as a kind of common language which can understand each other.

分类组建网络时,为了成功传输数据,发送方和接收方必须遵循用以交换信息的一套规则。这些计算机间交换数据的规则被称为协议。这套规则可以理解为一种彼此都能听得懂的公用语言。

Text (文本)

Protocols

There are two kinds of protocols about the network: "internal agreement" and "external agreement."

Internal agreement: In 1978, the International Organization for Standardization (ISO) set a standard for network communication Model, called the OSI/RM (Open System Interconnect/Reference Model). The structure is divided into seven layers, which from low to high respectively are the physical layer, data link layer, network layer, transport layer, session layer, presentation layer and application layer. Any network device of the upper and lower layer, has its specific agreement. At the same time, the same layer between the two devices, such as workstations and servers also has its agreement. Here, these agreements are defined as "internal agreement."

External protocol: External network formation must be selected by agreement. Because it is directly responsible for the computer to communicate with each other, external protocol is so often referred to as a network communication protocol. At first the development agreement of each company is to fit their own network communication, but with the popularity of network applications, requirements of interconnection between different networks have become more and more urgent, so the communication protocol has become the key technology to solve the problems between the network interconnections. Like people using a different mother language need a common language to talk with each other, network communication also needs a kind of common language, which is a communication protocol. At present, the local area network (LAN) that is commonly used in external protocols mainly includes NetBEUI, IPX/SPX and its compatible protocol and TCP/IP.

NetBEUI: NetBEUI agreement (NetBIOS Extended User Interface, User extension Interface) developed by IBM in 1985, is a small volume, highly efficient, high speed communicative protocol. The early product in Microsoft, such as DOS, LAN Manager, Windows 3.X and Windows for a Workgroup, chooses the NetBEUI as a communication protocol. Today, in the mainstream products of Microsoft, such as Windows XP and Windows NT, NetBEUI has become its inherent default protocol.

IPX/SPX: IPX/SPX (Internetwork Packet eXchange/Sequences Packet eXchange) is a set of communication protocols from Novell. The obvious difference between it and NetBEUI is that IPX/SPX is much larger, and that it has a strong adaptability in the complex environment. But in the non-Novell network environment, generally the IPX/SPX is not to be used.

TCP/IP: TCP/IP (Transmission Control Protocol/Internet Protocol Control Protocol/Internet Protocol) is one of the most commonly used communication protocols currently, which is a general Protocol in the computer world. In LAN, TCP/IP appeared on Unix systems at first, and now almost all the manufacturers and the operating systems support it. At the same time, it is also the foundation protocol of the Internet. TCP/IP has a high flexibility to support network of any size, and almost all of the servers and the workstations can be connected. But its flexibility has also brought a lot of inconvenience for using it, and the TCP/IP protocol needs to be set in complicated configurations before being used. Each node needs an "IP address," a "subnet mask," a "default gateway" and a "host name" at least. However, Windows NT provides a tool called dynamic host configuration protocol (DHCP), which can automatically distribute the information to the client for connecting the network, reduce the network burden, and avoid the error.

Key words and expressions (重点词汇)

receiver [rɪˈsiːvə] n. 接收器
exchange [ɪksˈtʃendʒ] n. 交换
protocol [ˈprəutəkɔl] n. 协议
architecture [ˈɑːkɪtektʃə] n. 架构
interconnection [ˌɪntəkəˈnekʃən] n. [计] 互连
compatible [kəmˈpætɪb(ə)l] adj. 兼容的
urgent [ˈəːdʒənt] adj. 紧急的
volume [ˈvɔljuːm] n. 大量
mainstream [ˈmeɪnstriːm] n. 主流
inherent [ɪnˈhɪrənt] adj. 固有的
manufacturer [ˌmænjuˈfæktʃərə] n. 制造商

flexibility [ˌfleksɪˈbɪlɪtɪ] n. 灵活性；适应性
configuration [kənˌfɪɡjəˈreʃən] n. 配置

Expanded vocabulary（扩展词汇）

reference model 参考模型
physical layer 物理层
data link laye 数据链路层
network layer 网络层
transport layer 传输层
session layer 会话层
presentation layer 表示层
application layer 应用层
referred to as 被称为
IP address IP 地址
subnet mask 子网掩码
default gateway 默认网关
host name 主机名

Oral practice（口语练习）

Q：How many kinds of protocols do you know?
A：Two kinds, they are "internal agreement" and "external agreement."
Q：What's the difference between them?
A：Internal agreements rarely involved in the network, they are mostly used for web developers. If only you want to form a network, just ignore the internal agreement.
Q：How many external protocols do you know?
A：External protocols mainly include NetBEUI, IPX/SPX and TCP/IP.

Exercises（练习题）

Ⅰ. Short-answer questions.

1. What is the protocol in computer?

2. Please list the seven layers of the OSI/RM.

3. How to set the configuration about TCP/IP?

4. Which communication protocol is widely used today?

Ⅱ. Translate the following paragraphs into Chinese.

1. When different types of microcomputers are connected in a network, the protocols can become very complex. Obviously, for the connections to work, these network protocols must adhere to certain standards.

2. The International Standards Organization has defined a set of communications protocols called the Open System Interconnection/Reference Model (OSI/RM). The purpose of the OSI/RM is to identify functions provided by any network. It separates each network's functions into seven "layers" of protocols, or communication rules.

Ⅲ. Translate the following paragraphs into English.

1. 如果要让两台实现互联的计算机间进行对话,它们两者使用的通信协议必须相同。否则中间还需要一个"翻译"进行不同协议的转换,这样不仅影响通信速度,而且不利于网络的安全和稳定运行。席卷全球的金融危机也在催生新的经济驱动力,物联网就是众人最为推崇的动力。

2. 如果网络中存在多个网段或要通过路由器相连时,就不能使用不具备路由器和跨网段操作功能的 NetBEUI 协议,而必须选择 IPX/SPX 或 TCP/IP 等协议。

协 议

网络中的协议有两类:"内部协议"和"外部协议"。

内部协议:1978年,国际标准化组织(ISO)为网络通信制定了一个标准模式,称为OSI/RM(Open System Interconnect/Reference Model,开放系统互联/参考模型)体系结构。该结构共分七层,从低到高分别是物理层、数据链路层、网络层、传输层、会话层、表示层和应用层。任何一个网络设备的上下层之间都有其特定的协议形式,同时在两个设备(如工作站与服务器)的同层之间也有协议约定。在这里,这些协议全部被定义为"内部协议"。

外部协议:外部协议即组网时所必须选择的协议。它由于直接负责计算机之间的相互通信,所以通常被称为网络通信协议。最初每家公司开发的协议,都是为了满足自己的网络通信,但随着网络应用的普及,不同网络之间进行互联的要求越来越迫切,因此通信协议就成为解决网络之间互联的关键技术。就像使用不同母语的人与人之间需要一种通用语言才能交谈一样,网络之间的通信也需要一种通用语言,这种通用语言就是通信协议。目前,局域网中常用的外部协议主要有NetBEUI、IPX/SPX及其兼容协议和TCP/IP三类。

NetBEUI协议:NetBEUI(NetBIOS Extended User Interface,用户扩展接口)由IBM于1985年开发完成,它是一种容量小、效率高、速度快的通信协议。微软在其早期产品,如DOS、LAN Manager、Windows 3.X和Windows for Workgroup中主要选择NetBEUI作为自己的通信协议。在微软如今的主流产品,如Windows XP和Windows NT中,NetBEUI已成为其固有的缺省协议。

IPX/SPX协议:IPX/SPX(Internetwork Packet eXchange/Sequences Packet eXchange,网际包交换/顺序包交换)是Novell公司的通信协议集。与NetBEUI的明显区别是,IPX/SPX显得更加庞大,在复杂环境下具有更强的适应性。但在非Novell网络环境中,一般不使用IPX/SPX。

TCP/IP协议:TCP/IP(Transmission Control Protocol/Internet Protocol,传输控制协议/网际协议)是目前最常用的一种通信协议,它是计算机世界里的一个通用协议。在局域网中,TCP/IP最早出现在Unix系统中,现在几乎所有的厂商和操作系统都支持它。同时,TCP/IP也是Internet的基础协议。TCP/IP具有很高的灵活性,支持任意规模的网络,几乎可连接所有的服务器和工作站。但其灵活性也给它的使用带来了许多不便,TCP/IP协议在使用时首先要进行复杂的设置。每个节点至少需要一个"IP地址"、一个"子网掩码"、一个"默认网关"和一个"主机名"。不过,在Windows NT中提供了一个称为动态主机配置协议(DHCP)的工具,它可自动为客户机分配连入网络时所需的信息,从而减轻了联网工作上的负担,也避免了出错。

Chapter 2

Network

2–6 Search Engines

Introduction（导读）

The Web can be an incredible resource providing information on nearly any topic imaginable. With over two billion pages and more being added daily, the Web is a massive collection of interrelated pages. With so much available information, locating the precise information you need can be difficult. Fortunately, a number of organizations called search services or search providers can help you locate the information you need. They maintain huge databases related to the information provided on the Web and the Internet. The information stored at these databases includes addresses, content descriptions or classifications, and keywords appearing on Web pages and other Internet informational resources. Additionally, search services provide special programs called search engines that you can use to locate specific information of the Web.

只要是你能想到的主题，你就能在难以置信的网络资源中找到相应的信息。每天有20多亿页资讯的更新，互联网成了汇聚众多相关信息的宝库。在如此之多的可获得的信息中，准确找到你想得到的信息是十分困难的。幸运的是一批被称为搜索服务和搜索提供者的组织为我们提供了网上信息定位服务。他们拥有涉及提供互联网信息的庞大的数据库。这些信息储存在这些数据库中，包括地址、内容概述或分类和网页上的关键字以及其他的网络信息资源。此外，搜索服务提供的搜索程序叫搜索引擎，我们能用它定位网络上的具体信息。

Text（文本）

Search Engines

Search engines are specialized programs that assist you in locating information on the Web and the Internet. To find information, you go to the searchservice's Web site and use their search engine. Search engine can provide many different search approaches.

1. **Keyword search engines**

In a keyword search, you enter a keyword or phrase reflecting the information you want. The search engine compares your entry against its database and returns a list of hits or sites that contain the keyword. Each hit includes a hyperlink to the referenced Web page along with a brief discussion of the information contained at that location. Many searches result in a large number of hits. For example, if you were to enter the keyword "Bayuquan," you would get over a thousand hits. Search engines order the hits according to those sites that most likely contain the information requested and present the list to you in that order.

2. Full text search engines

A full text search engine is a veritable search engine, and the representatives of foreign search engines are Google, Yahoo, etc. And the domestic well-known is Baidu, 360, and Youdao. They all extract information from the Internet site and establish the database, retrieve the records that match the conditions of the user, and then return the results to the user in a certain order.

From the point of search results, the full text search engines can be subdivided into two kinds; one is to rent other engines, such as Lycos, and the other is to have their own search programs, usually known as "spider" or "robot," and most search engines are used in this form.

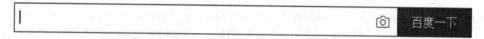

Baidu

3. Directory search engines

Most search engines also provide a directory or list of categories or topics such as Arts & Humanities, Business & Economics, Computers & Internet. In a directory search, also known as index search, you select a category that fits the information that you want. Another list of subtopics related to the topic you selected appears. You continue to narrow your search in this manner until a list of Web sites appears. This list corresponds to the hit list previously discussed. As a general rule, if you are searching for general information, use the directory search approach. For example, to find general information about text, use a directory search beginning with the category file. If you are searching for specific information, use the key word approach. For example, if you are looking for a specific docx file, use a key word search entering the file title or the file's name in the text selection box.

4. Metasearch engines

We visit several individual web sites to use search engines for researching a topic. At each site, enter the search instructions, wait for the hits to appear, review the list, and visit selected sites. This process can be quite time-consuming and duplicate responses from different search engines are inevitable. Metasearch engines offer an alternative.

When we enter the search instructions into metasearch engines, it will search the topic from many search engines at the same time, without searching the WWW by itself, and it does not have any database either. When users search a keyword, it converts the request to the command format of other search engines, and automatically submit your search request to several search engines simultaneously. The metasearch engine summarizes the receiving results, eliminates duplicates, orders the hits, and then provides the edited list to users' browsers. Using metasearch engines, the range of searching can be involved in multiple database engines. One of the best known Chinese metasearch engines is Souxing.

Key words and expressions（重点词汇）

precise [prɪˈsaɪs] *adj.* 精确的
engine [ˈendʒɪn] *n.* 引擎；发动机
specialize [ˈspeʃəlaɪz] *vi.* 专门从事
assist [əˈsɪst] *vt.* 帮助；促进
approach [əˈprəʊtʃ] *n.* 方法；途径
hit [hɪt] *n.* 热点
request [rɪˈkwest] *vt.* 要求；请求
veritable [ˈverɪtəb(ə)l] *adj.* 真正的；名副其实的
representative [reprɪˈzentətɪv] *n.* 代表；典型
extract [ˈekstrækt] *vt.* 提取
retrieve [rɪˈtriːv] *vt.* 检索
category [ˈkætɪɡ(ə)rɪ] *n.* 种类
correspond [kɔrɪˈspɔnd] *vi.* 符合；一致
duplicate [ˈdjuːplɪkeɪt] *adj.* 复制的；二重的
inevitable [ɪnˈevɪtəb(ə)l] *adj.* 必然的；不可避免的
alternative [ɔlˈtəːnətɪv] *n.* 供替代的选择

Expanded vocabulary（扩展词汇）

fortunately [ˈfɔːtʃənətlɪ] *adv.* 幸运地
organization [ˌɔːɡənaɪˈzeɪʃn] *n.* 组织；机构；体制；团体
phrase [freɪz] *n.* 短语
a list of 列表；清单
rent [rent] *vt.* 租用；租借
index [ˈɪndeks] *n.* 索引
submit [səbˈmɪt] *vi.* 提交
summarize [ˈsʌməraɪz] *vt.* 总结；概述
eliminate [ɪˈlɪmɪneɪt] *vt.* 消除

Oral practice（口语练习）

Q: How many search engines can you list?
A: Foreign search engines are Google, Yahoo, etc., and the domestic well-known is Baidu, 360, Youdao.

Q: Which do you often use?
A: Baidu is very popular, and I like to use it very much.
Q: What do you use it for?
A: I use it to search music and it can also help me in studying.

Exercises（练习题）

Ⅰ. Short-answer questions.

1. How many kinds of search engines can you list?

2. Try to talk about Keyword Search Engine.

3. How many kinds of full text search engines are there?

4. How did the metasearch engines work?

Ⅱ. Translate the following paragraphs into Chinese.

1. Are you planning a trip? Writing an Economics paper? Looking for a movie review? Trying to locate a long-lost friend? Information sources are related to these questions, and much, much more are available on the Web.

2. Special programs called agents, spiders, or robots continually look for new information and update the search services databases.

Ⅲ. Translate the following sentences into English.

1. 搜索引擎可以帮助我们在网上准确找到有用的信息。

2. 好的搜索引擎会根据用户查询的信息把查询结果按照一定的顺序组成列表，然后呈现给用户。

搜索引擎

搜索引擎是帮助你在互联网和因特网上定位信息的专用程序。你到搜索服务器网站用他们的搜索引擎去寻找信息。搜索引擎能提供许多不同的搜索方法。

1. 关键词搜索引擎

在关键词搜索引擎中，你输入一个能反映你想要的信息的关键词或者关键字。关键词搜索引擎会将它与整个数据库里的资源进行比较，然后呈现和关键词相匹配的列表和网站。每个热点包含一个带有简要介绍的超链接网站或者其他资源。例如，你输入一个关键词"鲅鱼圈"，你会获得上千条热点。搜索引擎根据你可能最需要的内容对网站进行排序，然后按照这个顺序将信息呈现出来。

2. 全文搜索引擎

全文搜索引擎是名副其实的搜索引擎，国外具有代表性的有谷歌、雅虎等，国内知名的有百度、360、有道等。它们都是从互联网上提取各个网站的信息而建立的数据库，检索与用户查询条件相匹配的相关记录，然后按一定的排列顺序将结果返回给用户。

从搜索结果的角度看，全文搜索引擎又可细分为两种：一种是租用其他引擎，如Lycos；另一种是拥有自己的搜索程序，俗称"蜘蛛"或者"机器人"，绝大部分搜索引擎都采用这种形式。（图略）

3. 目录式搜索引擎

多数的搜索引擎提供了主题分类列表，比如人文艺术、商业经济、计算机网络。在目录搜索中，也称检索搜索，你选择一个符合你想要的信息类别，其他的和主题相关的子类别列表会显示出来。以这种方式缩小你的搜索范围直到你想要的信息出现。这些目录列表是事先设计好的。通常查找普通的信息使用目录式搜索方法。例如，去找一个文本信息，在文件大类下搜索。如果你要找具体的信息，那么你就用关键词方式搜索。例如，如果你在找一个具体的 docx 文件，那么你可以在搜索框中输入文件的标题或者是文件的名字进行搜索。

4. 元搜索引擎

我们利用不同的独立站点的搜索引擎去研究某一主题。在每个站点都需要输入搜索指令，然后等待热点的出现，审查热点列表并访问所选择的网站。这个过程十分浪费时间，而且不同的搜索引擎返回相同的信息也是不可避免的。元搜索引擎给我们提供了另一种解决方案。

当元搜索引擎在接受用户查询请求时，它会同时在其他多个引擎上进行搜索。它自己不进行WWW的遍历，也没有自己的索引数据库。当用户查询一个关键词时，它把查询请求转换为其他搜索引擎的命令格式，并分别提交给其他搜索引擎。元搜索引擎汇总这些搜索引擎返回的结果，删除重复的内容，编辑后将结果返回给用户浏览器。利用元搜索引擎，查询范围可涉及多个引擎的数据库。著名的中文元搜索引擎之一是搜星。

Chapter 3

Software

3-1 Welcome to the New Office for Home—Office 365 Home

Introduction（导读）

From home to business, from desktop to web and the devices in between, Office delivers the tools to get work done. Office gives you the freedom to get work done virtually anytime, anywhere, and on any device.

从家庭到企业，从桌面到网页及相关设备，Office 提供理想工具，助你在任何时间、任何地点、任何设备上完成工作。

Text（文本）

Welcome to the New Office for Home—Office 365 Home

Office 365 Home：Office 365 Home is a subscription service built to help you and your family get things done from virtually anywhere and any device. What's included are as follows:

1. Full installed Office applications

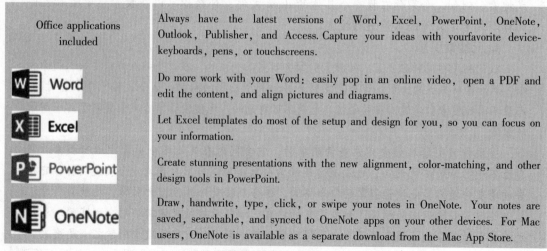

Office applications

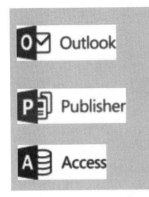

Outlook is now automatically connected to your Microsoft web mail account. Sync tasks and calendars with Outlook to put all your plans together. Mac users can sync only Exchange based e-mail accounts.

Work with pictures more easily in Publisher to add that eye-popping touch to your publications. Publisher is available for PC only.

Easily build a database app in Access and then use and share your app on the web. Access is available for PC only.

Office applications (continued)

2. Access across devices

Get the full installed Office experience on 5 PCs or Macs, 5 tablets (including iPad) and 5 phones. With the extra online storage Office 365 Home providing OneDrive, you always have your important documents and photos with you. And they're always up to date, whether you create, edit, or share them from your PC, Mac, iPad, Windows tablet, or smartphone. Office 365 and OneDrive keep your files in sync across your devices, and with just a click you can share files, even big ones, either by sending someone a link to a file or publishing it to Facebook.

3. Premium subscription benefits

1 TB of extra online storage per user, up to 5 users, and tech support at no extra charge. Office 365 services are being improved all the time and you get the latest improvements automatically with your subscription. As an Office 365 subscriber, you can get help right away from Microsoft-trained experts, by phone or chat at no extra charge.

Key words and expressions（重点词汇）

subscription [səbˈskrɪpʃn] n. 订阅；捐款
virtually [ˈvɜːtʃuəli] adv. 实际上；实质上
install [ɪnˈstɔːl] vt. 安装；安顿；安置；任命
application [ˌæplɪˈkeɪʃn] n. 应用；运用；申请；请求
include [ɪnˈkluːd] vt. 包括；包含
version [ˈvɜːʃən] n. 版本；译文
capture [ˈkæptʃə(r)] n. & vt. 捕捉；俘获；引起注意
content [ˈkɒntent] n. 内容；（书等的）目录；满足
align [əˈlaɪn] vt. & vi. 排列；使成一线；使结盟
diagram [ˈdaɪəɡræm] n. 图表；图解；示意图
template [ˈtempleɪt] n. 样板；模板；型板
focus on 致力于；予以注意
stun [stʌn] n. & vt. 使大吃一惊；打击；刺激
presentation [ˌpreznˈteɪʃn] n. 陈述；报告；介绍；赠送
swipe [swaɪp] vt. 刷（磁卡）；重击

sync ［sɪŋk］ *n. & v.* 同时；同步

account ［əˈkaʊnt］ *n.* 账；账目；存款；记述；报告

Expanded vocabulary（扩展词汇）

fuss ［fʌs］ *n.* 忙乱；大惊小怪

tablet ［ˈtæblət］ *n.* 碑；匾；药片；便笺簿；小块

extra ［ˈekstrə］ *adj.* 额外的；附加的

Skype ［skaip］ *n.* 网络电话

tech ［tek］ *n. & adj.* 技术；技术的

facebook ［ˈfeɪsbʊk］ *n.* 脸谱；美国一大学生网站

plus ［plʌs］ *prep.* 加；外加

mobile ［ˈməʊbaɪl］ *n. & adj.* 风铃；手机；可移动的

support ［səˈpɔːt］ *n. & vt.* 支持；帮助；支撑；支持者

charge ［tʃɑːdʒ］ *n. & vt.* 充电费用；装载；控诉

Oral practice（口语练习）

Q：What is your occupation?

A：Office secretary.

Q：What software do you use?

A：I use word-processing programs for my work.

Q：What would you do with the extra time?

A：To e-mail my friends with Outlook or make PowerPoint. Microsoft Office provides conveniences for us tremendously.

Exercises（练习题）

Ⅰ. Choices.

1. Which program can be used for dealing with characters?

   A. Word.　　　　B. Excel.　　　　C. PowerPoint.　　　　D. Access.

2. Which program can be used if you want to e-mail somebody?

   A. PowerPoint.　　B. Access.　　　C. Outlook.　　　　D. OneNote.

3. Which company released Office?

   A. Macromedia.　　　　　　　　B. Adobe.

   C. Microsoft.　　　　　　　　　D. Autodesk.

4. How many users can share the full installed Office 365 Home?

   A. 3.　　　　　B. 4.　　　　　C. 5.　　　　　D. 6.

5. How many applications does the Office 365 Home include?

   A. 3.　　　　　B. 5.　　　　　C. 7.　　　　　D. 9.

Chapter 3 Software

II. Translate the following sentences into Chinese.

1. From the first install, it indexes you based on your cell phone number.

2. An android version will also be launched.

3. Ask for content to be removed.

4. We isolated it earlier in another document, so just copy/paste in a new layer and align it.

5. Now you just swipe up from the camera icon.

III. Translate the following sentences into English.

1. 他的工作与计算机有关。

2. Office 办公软件是由微软公司发布的。

3. 大多数人一旦学会 Word，就会喜欢上它的。

4. 电子邮件已经逐渐替代了传统的信件。

5. 应用软件一定要经常使用才能熟练。

译文

欢迎使用新家庭版 Office——Office 365 家庭版

Office 365 家庭版：Office 365 家庭版是一项订阅服务，帮助你和你的家人随时随地在任何设备上完成工作。它包含以下内容：

1. 所有安装的 Office 应用程序

包含的 Office 应用程序	始终获得最新版本的 Word、Excel、PowerPoint、OneNote、Outlook、Publisher 和 Access。记录下你的奇思妙想——用你的键盘、钢笔或触摸屏。
Word	使用 Word 文档完成更多工作：轻松放入联机视频、打开 PDF 并编辑内容，以及对齐图片和图表。
Excel	让 Excel 模板为你完成大多数设置和设计工作，因此你可以全神贯注地处理你的信息。
PowerPoint	使用 PowerPoint 中新的对齐工具、颜色匹配工具和其他设计工具创建精美的演示文稿。
OneNote	在 OneNote 中绘制、手写、键入、单击或轻扫你的笔记。你的笔记将被保存下来，可供搜索，且会同步到其他设备上的 OneNote 应用程序。Mac 用户可从 Mac 应用商店单独下载 OneNote。
Outlook	Outlook 现在可以自动连接到你的 Microsoft Web 电子邮件账户。使用 Outlook 同步任务和日历，将所有计划整合在一起。Mac 用户只能同步基于 Exchange 的电子邮件账户。
Publisher	在 Publisher 中更轻松地处理图片，为你的出版物增添惊艳效果。Publisher 仅向 PC 提供。
Access	在 Access 中轻松构建数据库应用程序，然后在 Web 上使用和共享你的应用程序。Access 仅向 PC 提供。

Office 应用程序

2. 跨设备访问

在 5 台 PC 或 Mac、5 台平板电脑（包括 iPad）和 5 部手机上畅享完整安装的 Office 体验。凭借 Office 365 家庭版通过 OneDrive 提供的额外的网络存储空间，你的重要文档和照片始终如影随形。无论你是在 PC、Mac、iPad、Windows 平板电脑还是在智能手机上创建、编辑或共享，这些文档和照片始终是最新的。Office 365 和 OneDrive 让你的文件在不同的设备上保持同步。只需点击一下发送文件链接或发布至 Facebook，你就可以共享文件，甚至是很大的文件。

3. 高级版订阅权益

最多 5 名用户，每名用户可享用 1 TB 额外的联机存储空间。你的订阅还包括免费技术支持等其他服务。Office 365 服务不断改进。通过订阅，你可以自动获得最新改进。作为 Office 365 的订阅用户，无须额外付费，你即可以通过电话或聊天工具立即获得经过微软培训的专业人员的帮助。

Chapter 3

Software

3-2 What Is Photoshop?

Introduction（导读）

Have you dealt with any photos? In this task we will study Photoshop! You should know what Photoshop is, Photoshop's developing history and the English version of the Photoshop interface.

你处理过照片吗？在这个任务中我们将学习 Photoshop! 你将会知道 Photoshop 是什么、Photoshop 的发展历史以及英文版的界面。

Text（文本）

What Is Photoshop?

Photoshop is one of the most famous image processing softwares of Adobe, a wonderful software which integrates many image processing functions, such as scanning, modification, image producing, advertising creation and image input and output. That is why it is favored by the majority of graphic designers and computer art lovers.

Photoshop was created in 1988 by Thomas and John Knoll. Since then, it has become the de facto industry standard in raster graphics editing, such terms as "photo shopping" and "Photoshop contest" were born. It can edit and compose raster images in multiple layers and support masks, alpha compositing and several color models including RGB, CMYK, Lab color space, spot color and duotone. Photoshop has vast support for graphic file formats but also uses its own PSD and PSB file formats which support all the aforementioned features. In addition to raster graphics, it has limited abilities to edit or render text, vector graphics, 3D graphics and videos. Photoshop's feature set can be expanded by Photoshop plug-ins, programs developed and distributed independently of Photoshop that can run inside it and offer new or enhanced features.

Photoshop's naming scheme was initially based on version numbers. However, on October 2003, following the introduction of Creative Suite branding, each new version of Photoshop was designated with "CS" plus a number, e.g. the eighth major version of Photoshop was Photoshop CS and the ninth major version was Photoshop CS2. Photoshop CS3 through CS6 were also distributed in

Chapter 3 Software

Thomas Knoll   **John Knoll**

two different editions: Standard and Extended. In June 2013, with the introduction of Creative Cloud branding, Photoshop's licensing scheme was changed to that of software as a service and the "CS" suffixes were replaced by "CC". Historically, Photoshop was bundled with additional software such as Adobe Image Ready, Adobe Fireworks, Adobe Bridge, Adobe Device Central and Adobe Camera RAW.

The following picture is the English version of the Photoshop interface.

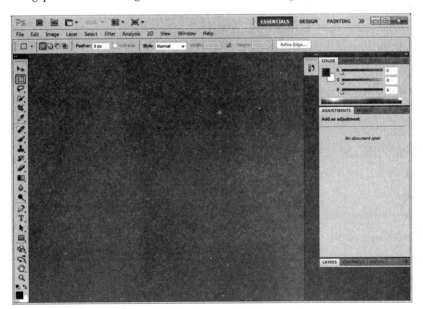

Photoshop interface

Key words and expressions (重点词汇)

software ['sɔftwɛə] n. 软件
scan [skæn] vt. 扫描
advertise ['ædvətaiz] vt. 做广告
integrate ['ɪntigreit] vt. 使一体化；使整合；使完整；使结合成为整体
term [tə:m] n. 术语
contest ['kɔntest] vt. 竞赛

compose ［kəm'pəuz］ vt. 组成，制作
duotone ［'djuːətəun］ n. 双色调
vast ［vɑːst］ adj. 广阔的
format ［'fɔːmæt］ n. 格式
render ［'rendə］ vt. 渲染
expand ［ik'spænd］ vt. 扩展
distribute ［di'stribjuːt］ vt. 发布
scheme ［skiːm］ n. 设计
license ［'laisəns］ n. 许可
suffix ［'sʌfiks］ n. 后缀
historically ［his'tɔrikli］ adv. 从历史角度；在历史上

Expanded vocabulary（扩展词汇）

image processing 图像处理
graphic designers 美术设计员
de facto industry standard 事实上的工业标准
raster graphics editing 光栅图形编辑
multiple layers 多层
aforementioned features 上述功能
Creative Suite branding　Creative Suite 品牌

Oral practice（口语练习）

Teacher：Let's open the Photoshop software.
Student：OK.
Teacher：The interface includes six parts. The main parts are tools panel, application bar and Photoshop panels.
Student：I see.
Teacher：Can you know where they are?
Student：No, I don't.
Teacher：This is not a problem. I can tell you. Here is the application bar. This is the tools panel. That is the Photoshop panels.
Student：Thanks, teacher！I know. They are so easy.

Exercises（练习题）

Ⅰ. Choices.

1. Photoshop is one of the most famous _____ processing software.
   A. image　　B. creative　　C. program　　D. expand
2. Photoshop is _____ by the majority of graphic designers and computer art lovers.

  A. favore B. favored C. favire D. favorite
3. Photoshop uses its own _____ and PSB file formats.
  A. jpg B. bmp C. PSD D. image
4. Photoshop can edit and compose raster images in _____ layers.
  A. video B. multiple C. multi D. multiples
5. Photoshop's feature set can be _____ by Photoshop plug-ins.
  A. expensive B. expande C. expand D. expanded

Ⅱ. Translate the following sentences into Chinese.

1. Photoshop's naming scheme was initially based on version numbers.

2. Photoshop is a wonderful software which integrates many image processing functions.

3. Photoshop was created in 1988.

4. Photoshop is favored by the majority of computer art lovers.

5. Photoshop uses its own file formats.

Ⅲ. Translate the following sentences into English.

1. Photoshop 能做图形处理。

2. Photoshop 是最常用的软件。

3. Photoshop 的版本很多。

4. Photoshop 的界面很简单。

5. 我非常爱使用 Photoshop。

什么是 Photoshop？

Photoshop 图像处理软件是 Adobe 公司最著名的图像处理软件之一，是一个精彩的、集成了多种图像处理功能的软件，如扫描、修改、图像制作、广告创意和图像的输入和输出，这就是它深受多数平面设计师和电脑美术爱好者喜爱的原因。

Photoshop 图像处理软件是由托马斯和约翰·诺尔在 1988 年创建的。自那时以来，它已成为光栅图形编辑事实上的行业标准，因此像"照片商店"和"Photoshop 竞赛"这样的术语也随之出世。它可以在多个层面上编辑和制作光栅图像，支持遮罩、alpha 合成及几种颜色模式，包括 RGB、CMYK、LAB 色彩空间、专色和双色调。Photoshop 图像处理软件为图形文件格式提供广泛的支持而且也使用它自己的 PSD 和 PSB 文件格式支持所有上述功能。除了光栅图形，它具有有限的能力来编辑或渲染文本、矢量图形、三维图形和视频。Photoshop 图像处理软件的功能可以通过 Photoshop 图像处理软件插件得以扩展，这些插件是开发的、独立于 Photoshop 的程序，它们可以在 Photoshop 程序里面运行并提供新的或增强的功能。

Thomas Knoll

John Knoll

Photoshop 的命名方案最初是基于版本号。然而，在 2003 年 10 月，Creative Suite 品牌被推出之后，Photoshop 的每个新版本都标有"CS"和一个数字，如 Photoshop 图像处理软件的第八个主要版本是 Photoshop CS，第九个主要版本是 Photoshop CS2。Photoshop CS3 到 CS6 也发布了两个不同的版本：标准版和扩展版。2013 年 6 月，随着创意云品牌的推出，Photoshop 图像处理软件的许可计划被改为软件服务，同时

"CS"的后缀被改为"CC"。从历史上看，Photoshop是和多个软件捆绑在一起的，如Adobe Image Ready、Adobe Fireworks、Adobe Bridge、Adobe Device Central和Adobe Camera RAW。

下面的图片是Photoshop界面的英文版。

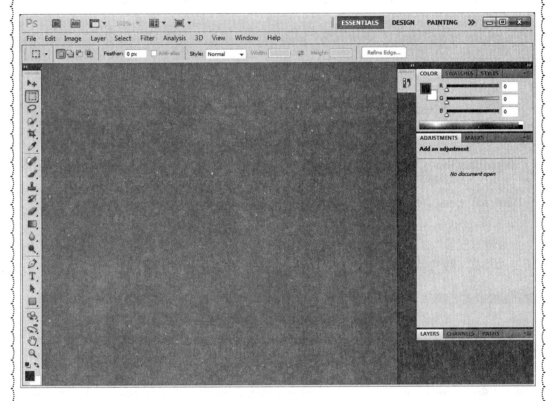

Photoshop 界面

Chapter 3

Software

3-3 What Is Adobe Premiere Pro?

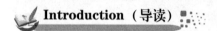
Introduction（导读）

Have you made any videos? In this task you should know what Adobe Premiere Pro is, premiere's developing history and the English version of the Adobe Premiere Pro interface.

你处理过视频吗？在这个任务中你将会知道 Adobe Premiere Pro 是什么、Premiere 的发展历史以及英文版的界面。

Text（文本）

What Is Adobe Premiere Pro?

Adobe Premiere Pro is a timeline-based video editing software application. It is part of the Adobe Creative Cloud, which includes video editing, graphic design, and web development programs.

Premiere Pro is used by broadcasters such as the BBC and CNN. It has been used to edit feature films, such as "Gone Girl," "Captain Abu Raed," and "Monsters", and other venues such as "Madonna's Confessions Tour."

Premiere Pro is the redesigned successor to Adobe Premiere, and was launched in 2003. Premiere Pro refers to versions released in 2003 and later, whereas Premiere refers to the earlier releases. Premiere was one of the first computer-based NLES (non-linear editing system), with its first release on Mac in 1991. Up until version Premiere Pro 2.0 (CS2), the software packaging featured a galloping horse, in a nod to Eadweard Muybridge's work, "Sallie Gardner at a Gallop".

Premiere Pro supports high resolution video editing at up to 10,240 × 8,192 resolution, at up to 32-bits per channel color, in both RGB and YUV. Audio sample-level editing, VST audio plug-in support, and 5.1 surround sound mixing are available. Premiere Pro's plug-in architecture enables it to import and export formats beyond those supported by QuickTime or DirectShow, supporting a wide variety of video and audio file formats and codes on both MacOS and Windows. When used with Cineform's Neo line of plug-ins, it supports 3D editing with the ability to view 3D material using 2D monitors, while making individual left and right eye adjustments.

The following picture is the English version of the Adobe Premiere Pro interface.

Adobe Premiere Pro interface

Key words and expressions（重点词汇）

software ['sɔftwɛə] *n.* 软件
application [æpli'keiʃn] *n.* 应用
include [in'kluːd] *vt.* 包括；包含
broadcaster ['brɔːdkɑːstə] *n.* 广播员；广播公司
monster ['mɔːnstə] *n.* 怪物；恶魔 *adj.* 巨大的；庞大的
venue ['venjuː] *n.* 犯罪地点；案发地点；会场；（尤指）体育比赛场所；审判地
successor [sək'sesə] *n.* 接替的人或事物；继承人；继任
launch [lɔːnt] *vt.* 发射；[计算机] 开始（应用程序）；发动；开展（活动、计划等）
release [ri'liːs] *vt.* 释放；放开；发布；发行 *n.* 释放；排放
whereas [weər'æz] *conj.* 鉴于；然而；反之
feature ['fiːtʃə] *vt.* 使有特色；描写……的特征；以……为号召物 *n.* 特征；特点
galloping ['gæləpɪŋ] *adj.* 飞驰的；急性的 *v.* （使马）飞奔；奔驰（gallop 的现在分词）；快速做 [说] 某事
resolution [rezə'luːʃn] *n.* 分辨率；决心；解决；坚决
architecture ['ɑːkitektʃə] *n.* 建筑学；建筑风格；体系结构；（总体、层次）结构
variety [və'raiəti] *n.* 多样；种类；杂耍；变化；多样化

Expanded vocabulary（扩展词汇）

timeline-based 基于时间线的

graphic design 平面设计
feature films 故事片；长片
Captain Abu Raed《阿布拉德机长》
Madonna's Confessions Tour《麦当娜忏悔之旅》
NLES（non-linear editing system）非线性编辑系统
in a nod to 在点头，对……表示肯定
Eadweard Muybridge 埃德沃德·迈布里奇（拍摄动作细节的摄影家）
Sallie Gardner at a Gallop《飞驰中的萨利·加德纳》（1880 年的美国纪录片）
sample-level 样本级别
VST 虚拟演播技术
5.1 surround sound 5.1 环绕声
DirectShow 微软流媒体处理的开发包
Cineform's Neo line of plug-in Cineform 的新线插件

Oral practice（口语练习）

Teacher：Let's open the Premiere Pro software.
Student：OK.
Teacher：The interface includes six parts. The main parts are project window, application bar and time line window.
Student：I see.
Teacher：Do you know where they are?
Student：No, I don't.
Teacher：This is not a problem. I can tell you. Here is the application bar. Thist is the project window. That is the time line window.
Student：Thanks, teacher! I know. They are so easy.

Exercises（练习题）

Ⅰ. Choices.

1. Adobe Premiere Pro is a timeline-based video editing software _____ .
    A. application   B. editor   C. program   D. maker
2. Premiere Pro is _____ by broadcasters such as the BBC and the CNN.
    A. use   B. uses   C. used   D. user
3. Premiere Pro was _____ in 2003.
    A. release   B. released   C. releases   D. releasing
4. Premiere was one of the _____ computer-based NLES.
    A. second   B. one   C. first   D. third
5. The software packaging of Premiere Pro 2.0 featured a _____ horse.
    A. galloped   B. gallop   C. gallopers   D. galloping

Chapter 3 Software

II. Translate the following sentences into Chinese.

1. Premiere Pro supports high resolution video editing.

2. Premiere refers to the earlier releases.

3. Premiere Pro refers to versions released in 2003.

4. Premiere Pro is the redesigned successor to Adobe Premiere.

5. Premiere has been used to edit feature films.

III. Translate the following sentences into English.

1. Premiere Pro 能做什么呢？

2. Premiere Pro 支持多种视频编辑。

3. Premiere Pro 有很多插件。

4. Premiere Pro 的界面很简单。

5. 我非常了解 Premiere Pro。

什么是 Adobe Premiere Pro？

Adobe Premiere Pro 是一款编辑基于时间轴视频的应用软件。这是 Adobe Creative 云的一部分，其中包括视频编辑、平面设计、网站开发项目。

Premiere Pro 是由广播公司如英国广播公司和美国有线电视新闻网使用。它已被用于编辑故事片，如《消失的女孩》、《阿布拉德机长》、《怪兽》以及其他影片如《麦当娜忏悔之旅》。

Premiere Pro 是 Adobe Premiere 的替代品，是 2003 年发布的。Premiere Pro 指的是 2003 年及以后发布的版本，而 Premiere 指的是早期发布的版本。Premiere 是最早基于非线性编辑的计算机系统之一，它于 1991 年第一次发布在 Mac 上。直到 Premiere Pro 2.0（CS2）版本，这个版本的软件包以一匹奔跑的马为特征，这是对埃德沃德·迈布里奇的作品《飞驰中的萨利·加德纳》的一种肯定。

Premiere Pro 支持最高可达 10 240×8 192 高分辨率的视频编辑，在 RGB 和 YUV 中，最高可支持每通道 32 位颜色。音频采样级别的编辑、虚拟演播技术音频插件的支持以及 5.1 环绕声的混合也是可用的。Premiere Pro 的插件结构使得它支持的输入和输出格式远超 QuickTime 或 DirectShow，支持 MacOS 和 Windows 上大量的、多种多样的视频和音频文件格式和解码器。当使用 Cineform 新线插件时，它支持 3D 编辑，当单独使用左眼或右眼调整时，它可以使用 2D 显示器查看 3D 素材。

下面的图片是 Adobe Premiere Pro 界面的英文版。

Adobe Premiere Pro 界面

Chapter 3
Software

3-4 Computer Language—Communication Tool Between Human and Computer

Introduction（导读）

How does a computer work? In fact, a single action or a step of computers is run in accordance with programs written by computer languages in advance. In order to control computers well, people must send commands to the computer through the computer language.

计算机是如何工作的？实际上，计算机每做的一次动作、一个步骤，都是按照事先用计算机语言编写好的程序来执行的。为了更好地控制计算机，人们必须通过计算机语言向计算机发出命令。

Text（文本）

Computer Language—Communication Tool Between Human and Computer

Computer language, also known as programming language, is the communication tool between human and computer. It is usually used to write a program that can be run by a computer. In general, computer language can be divided into three kinds: machine language, assembly language and high-level language. Machine language and assembly language belong to low-level language.

1. Machine language

Machine language is a set of binary instructions performed directly by a computer's central processing unit. Each different type of central processing unit has its own machine language, that is, the language is machine-dependent. The computing efficiency of machine language is

Kinds of computer languages

the highest in all languages, but machine language programs are not portable to other computers and hard to remember. Computer programs are rarely written directly in machine language.

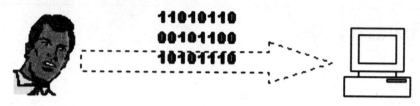

Machine language

2. Assembly language

To make it easier for programmers to write programs, assembly language was invented soon. It maps machine instructions to human-readable mnemonics. For example:

```
mov   ax, x   ; put x into ax register
add   ax, y   ; add y to the ax register
```

Assembly language

In this case the ax register is a special memory location. The mnemonic codes are "mov" and "add"; "mov" is short for move, and "add" is the English word "add." Assembly language is often difficult to learn and is machine-dependent like machine language.

3. High-level language

High-level language allows programmers to write code that is closer to the way we think of a problem. For example:

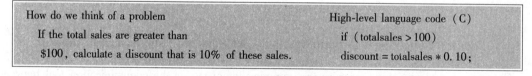

How do we think of a problem	High-level language code (C)
If the total sales are greater than	if (totalsales > 100)
$100, calculate a discount that is 10% of these sales.	discount = totalsales * 0.10;

High-level language

The main advantage of high-level language over low-level language is that it is easier to read, write, maintain and not machine-dependent. Today's favoured languages are C, C++, C#, Java, HTML, Visual C++, Lisp, Prolog, etc.

As machine language is the only language that computer can recognize, programs written in a high-level language can't be run until they are translated into machine language by a compiler or interpreter.

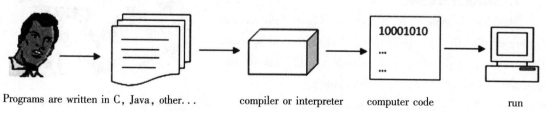

Programs are written in C, Java, other...   compiler or interpreter   computer code   run

The execution of high-level language

Chapter 3　Software

Key words and expressions（重点词汇）

programming ['prəʊɡræmɪŋ] *n.* [计] 程序设计；编程；利用 program 同根词构成新词
　　　　　　　（如 programmer 程序员；编程员）
program ['prəʊɡræm] *n.* 节目；节目单；计划；安排；[计] 程序；编程；用编程语
　　　　　　　言编写的可实现一定功能的代码
machine language [计] 机器语言
machine-dependent [mə'ʃiːndɪ'pendənt] *adj.* 依赖于机器的；与机器有关的
high-level language [计] 高级语言
binary ['baɪnəri] *adj.* 二态的；二元的；双重的；[计] 二进制的
　　　　　　　（二进制数据是用 0 和 1 两个数码来表示的数）
efficiency [ɪ'fɪʃnsi] *n.* 效率；效能；实力；能力
rarely ['reəli] *adv.* 很少地；罕有地；极精彩地；珍奇地；绝佳地
programmer ['prəʊɡræmə(r)] *n.* [计] 程序员；编程员
human-readable [h'juːmən'riːdəbl] *adj.* 可读的；人类可读的
register ['redʒɪstə(r)] *n.* [计] 寄存器；注册；登记
maintain [meɪn'teɪn] *vt.* 维护；保持；保养；坚持；固执己见

Expanded vocabulary（扩展词汇）

in accordance with 根据；与……一致；依照
assembly [ə'sembli] *n.* 装配；集会；集合
assembly language [计] 汇编语言
instruction [ɪn'strʌkʃn] *n.* [计] 指令；指示；指导
portable ['pɔːtəbl] *adj.* [计] 可移植的；手提的；便携式的
mnemonic [nɪ'mɒnɪk] *n. & adj.* [计] 助记符；记忆的；记忆术的
compiler [kəm'paɪlə(r)] *n.* [计] 编译器；汇编者；编辑者；编纂者
interpreter [ɪn'tɜːprɪtə(r)] *n.* 翻译；解释器模式；[计] 解释器；解释程序（利用
　　　　　　　interpret 同根词构成新词，如 interpretable 可解释的、可
　　　　　　　翻译的；[计] 指在计算机专业中对英语单词的解释）

Oral practice（口语练习）

Q：How do we communicate with computer?
A：By way of computer languages.
Q：What are the commonly used languages?
A：C, Java, Visual Basic...
Q：Could you give me an example of application?
A：OK. What to do first after buying a new computer?
Q：Install the operating system such as Windows XP, Windows 7, etc.
A：The Windows operating system is written by C, C++, assembly language, etc.

Q: Where does Java apply to?

A: Network programming and development of the mobile phone, like mobile phone games.

Q: But computer only recognizes 0 and 1.

A: So programs can't be run until they are translated into machine language.

Exercises (练习题)

Ⅰ. Choices.

1. _____ is usually used to write a program that can be run by a computer.
   A. Human language   B. Computer Language
   C. Tool   D. Compiler

2. The following languages are computer languages except _____.
   A. machine language   B. assembly language
   C. English   D. high-level language

3. The computing efficiency of machine language is _____ in all languages.
   A. the highest   B. the lowest
   C. lower   D. higher

4. _____ maps machine instructions to human-readable mnemonics.
   A. Machine language   B. High-level language
   C. Assembly language   D. Low-level language

5. High-level languages are easier to _____ and not machine-dependent.
   A. write   B. read
   C. maintain   D. all the above answers are right

Ⅱ. Translate the following sentences into Chinese.

1. Finding a programmer to write the computer program isn't a problem.

2. Most programs that can be written in a high-level language are processor-independent.

3. In machine languages, instructions are written as sequences of 1s and 0s, called bits, which a computer can understand directly.

4. Programming languages allow larger and more complicated programs to be developed faster.

5. The first assembly languages emerged in the late 1950s with the introduction of commercial computers.

Ⅲ. Translate the following sentences into English.

1. Java 语言是 Sun 公司在 1991 年开发的。

2. 一个程序是一系列的指令，它告诉计算机执行哪些操作。

3. 事实上，早期的程序是直接用机器语言写的。

4. 高级语言是比机器代码或汇编语言更接近于自然语言或数学语言的一种语言。

5. 编程语言几乎可以追溯到 20 世纪 40 年代数字计算机发明之时。

计算机语言——人类和计算机之间的交流工具

计算机语言，也被称为编程语言，是人与计算机之间通信的工具，它通常是用来编写可由计算机运行的程序。总的来说，计算机语言可以分成三类：机器语言、汇编语言和高级语言。机器语言和汇编语言属于低级语言。

1. 机器语言

机器语言是计算机的中央处理器（CPU）能直接执行的一系列二进制指令集合，不同类型的中央处理器有它自己的机器语言，也就是说，这种语言是依赖于机器的。机器语言的运算效率是所有语言中最高的，但是由于机器语言程序无法移植到其他计算机，并难于记忆，计算机程序很少直接用机器语言来书写。

各种计算机语言

机器语言

2. 汇编语言

为了使程序员编写的程序更容易，不久发明了汇编语言。它是把机器指令变换成易于阅读的助记符，例如：

```
mov   ax, x   ；把 x 放入 ax 寄存器中
add   ax, y   ；把 y 与 ax 寄存器中数值的和放入 ax 寄存器中
```

汇编语言

在这个例子中 ax 寄存器是一个专门的存储器，助记符是"mov"和"add"，"mov"是"移动"的简写，"add"是"加"的英文单词。和机器语言一样，汇编语言往往是很难学习并且也是依赖于机器的语言。

3. 高级语言

高级语言允许程序员以更接近于我们思考问题的方式编写代码。例如：

```
我们如何思考问题                    高级语言代码（以 C 语言为例）
如果销售总额大于 100 美元，          if（totalsales > 100）
    计算折扣为销售总额的百分之十。        discount = totalsales * 0.10；
```

高级语言

高级语言与低级语言相比，主要优势是很容易阅读、书写、维护并且不依赖于机器，现在常用的语言有 C、C++、C#、Java、HTML、Visual C++、Lisp、Prolog 等。

由于机器语言是计算机唯一识别的语言，使用高级语言编写的程序要等到它们被编译程序或解释程序翻译成机器语言才能运行。

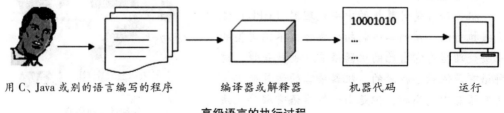

用 C、Java 或别的语言编写的程序　　　编译器或解释器　　　机器代码　　　运行

高级语言的执行过程

Chapter 3

Software

3-5 Application Fields of Database

Introduction（导读）

A database is a collection of logically related data elements that is organized so that its contents can easily be accessed, managed, and updated. The common database management system has Access, SQL Server, Mysql, Oracle, etc. What are the application fields of a database? You will know the answer after reading this article.

数据库是一个逻辑相关数据元素的集合，按照容易存取、管理及更新的原则进行组织。常见的数据库管理系统有 Access，SQL Server，Mysql，Oracle 等。数据库有哪些应用领域？阅读完这篇文章，你就会知道答案。

Text（文本）

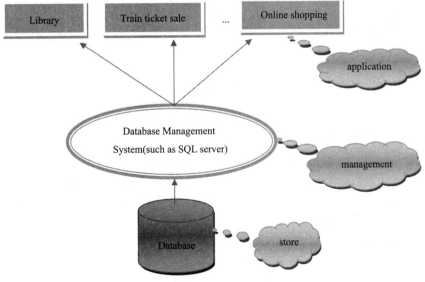

Application examples of database

The application fields of a database are very wide. Families, companies, a large enterprise, or government departments all need to use a database to store data information. A large part of traditional databases is used in the fields of business, such as the securities industry, bank, the sales department, hospital, company or business unit, and the government departments, national defense and military fields, science and technology development area, etc.

With the development of information age, the database also produces some new application fields mainly in the following six aspects.

1. Multimedia database

This kind of database mainly stores data related to multimedia, such as sound, image and video data. The main feature of multimedia data is that data is continuous; the data quantity is large, and huge storage space is required.

2. Mobile database

The development of mobile database is based on the mobile computer system, such as laptop, palmtop, etc. The biggest feature of this database is transmission through the wireless digital communication network. Mobile database can capture and access data whenever and wherever possible, and bring great convenience for some business applications and some emergency situations.

3. Space database

Space database includes geographic information systems (GIS), and computer aided design (CAD). The purpose is to effectively use satellite remote sensing resources to draw various economic topic maps quickly.

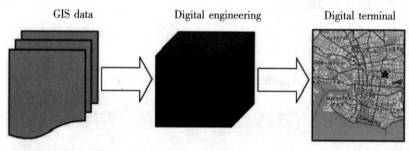

Digital community, digital city, digital earth...
Digital protection zone, digital ecological station...

Application of space database

4. Information retrieval system

Information retrieval system sets up a procedural system about information collection, processing, storage and retrieval according to the specific information needs, and its main purpose is to provide information service for people. Search website commonly used in daily life has Baidu, Yahoo, Sohu, etc.

5. Distributed database

Distributed database system is suitable for decentralized departments. It allows each department to store commonly used data in the local computer, and the majority of processing data is finished in

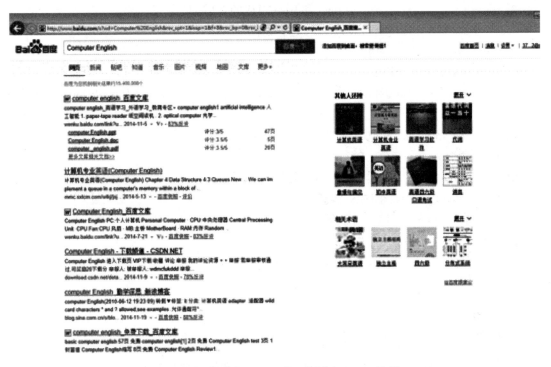

The result of search "Computer English" in www. Baidu. com

the local computer. Different computers are connected by the data communication network.

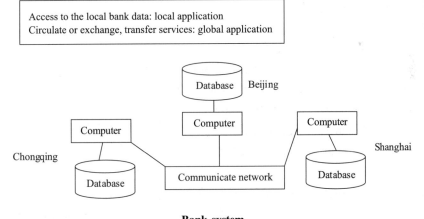

Bank system

6. Expert decision system

Expert decision system is a computer program with a large number of specialized knowledge, and use the knowledge to solve practical problems of a specific field, such as diagnosis system (medical care, fault), forecast system (weather, traffic), control system (battlefield control).

Key words and expressions （重点词汇）

database ['deɪtəbeɪs] n. 资料库；信息库；［计］数据库

（由 data 和 base 组成的合成词）

access ['ækses] *vt.* ［计］存取；接近；访问

update [ˌʌp'deɪt] *vt.* 更新；修改

application field 应用领域

multimedia [ˌmʌlti'miːdiə] *n.* 多媒体（multi 为前缀，表示许多，如 multi-processor 多核处理器）

laptop ['læptɔp] *n.* 笔记本电脑；便携式电脑

feature ['fiːtʃə] *n.* 特色；特征

transmission [træns'mɪʃn] *n.* 传输；传送；发射

capture ['kæptʃə] *vt.* 捕捉；俘获

retrieval [rɪ'triːvl] *n.* 检索；恢复；取回；拯救

procedural [prə'siːdʒərəl] *adj.* 程序的；有关程序的

website ['websaɪt] *n.* ［计］网站（web 为前缀，表示"网络"，如 webpage 网页，webmaster 网管）

be suitable for 适于

Expanded vocabulary（扩展词汇）

logically ['lɔdʒɪklɪ] *adv.* 逻辑上；符合逻辑地；能推理地

related [rɪ'leɪtɪd] *adj.* 相关的；有关系的

traditional [trə'dɪʃənl] *adj.* 传统的；口传的；惯例的；因袭的

securities industry 证券行业

national defense 国防

military ['mɪlətri] *adj.* 军事的；军用的

continuous [kən'tɪnjuəs] *adj.* 连续的；延伸的；持续的

palmtop ['pɑːmtɔp] *n.* ［计］掌上计算机

geographic information system 地理信息系统

computer aided design 计算机辅助设计

distributed [dɪs'trɪbjuːtɪd] *adj.* 分布式的

circulate or exchange 通兑

diagnosis [ˌdaɪəg'nəusɪs] *n.* 诊断；诊断结论；判断；结论

（［计］指在计算机专业中对英语单词的解释）

 Oral practice（口语练习）

Q：What do you like to do in free time?

A：Play games, go shopping, read in library, etc.

Q：Where is the data stored and how is the data processed?

A：I don't know.

Q：A database.

A：What is a database?

Chapter 3　Software

Q: Store and manage data in computer.
A: What is the common database software?
Q: Access, SQL Server.
A: Does data refer to the digital only?
Q: No, it also includes text, graphics, images, voice.
A: Oh, database is very useful.

Exercises (练习题)

Ⅰ. Choices.

1. The following system is database management system except _____.
    A. SQL Server   B. Access
    C. Java   D. Oracle

2. _____ the development of information age, the database also produces some new application fields.
    A. On   B. Under
    C. For   D. With

3. Mobile database can capture and access data whenever and wherever possible, _____ bring great convenience for some business applications and some emergency situations.
    A. as   B. but
    C. and   D. so

4. Space database includes _____.
    A. CAD and GIS   B. DB and CAI
    C. GIS and DB   D. CAI and GIS

5. Distributed database allows each department to store commonly used data in the local computer, the majority of _____ is finished in the local computer.
    A. search data   B. processing data
    C. insert data   D. delete data

Ⅱ. Translate the following sentences into Chinese.

1. Databases can be stored on magnetic disk or tape, optical disk, or some other secondary storage devices.

2. The Google comes from "googol," the number 1 followed by 100 zeros. The name is also used as a verb; for example, "to Google something" means to search the Web for it.

3. An effective database management system lets you retrieve only the information you need for a specific purpose.

4. Data security prevents unauthorized users from viewing or updating the database. Using passwords, users are allowed access to the entire database or subsets of the database.

5. SQL stands for Structured Query Language.

Ⅲ. Translate the following sentences into English.

1. 一个数据库由一个文件或文件集合组成。

2. 数据库可以用于存储各种数据,如数字、声音、图像。

3. SQL Server 是一个客户机/服务器模式的关系型数据库管理系统。

4. 在过去,一个机构中的每个应用程序都使用自己的文件。

5. 数据库管理系统定义、创建和维护数据库。

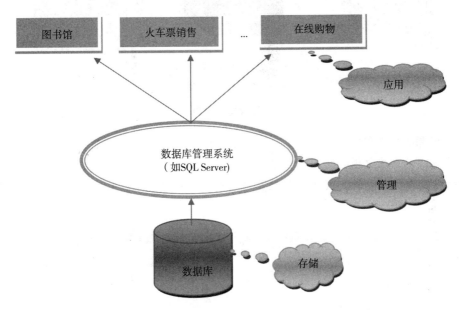

数据库的应用实例

数据库应用领域非常广泛，不管是家庭、公司或大型企业，还是政府部门，都需要使用数据库来存储数据信息。传统数据库中的很大一部分用于商务领域，如证券行业、银行、销售部门、医院、公司或企业单位，以及国家政府部门、国防军工领域、科技发展领域等。

随着信息时代的发展，数据库也相应产生了一些新的应用领域，主要表现在以下六个方面。

1. 多媒体数据库

这类数据库主要存储与多媒体相关的数据，如声音、图像和视频等数据。多媒体数据最大的特点是数据连续，而且数据量比较大，存储需要的空间巨大。

2. 移动数据库

移动数据库是在移动计算机系统上发展起来的，如笔记本电脑、掌上计算机等。该数据库最大的特点是通过无线数字通信网络传输。移动数据库可以随时随地获取和访问数据，为一些商务应用和一些紧急情况带来了很大的便利。

3. 空间数据库

空间数据库主要包括地理信息系统（即 GIS）和计算机辅助设计（即 CAD）。其目的是有效地利用卫星遥感资源迅速绘制出各种经济专题地图。

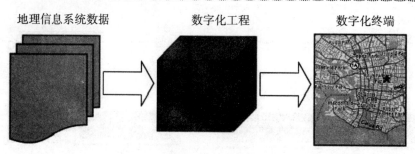

空间数据库应用

4. 信息检索系统

信息检索系统是根据特定的信息需求而建立起来的一种有关信息搜集、加工、存储和检索的程序化系统，其主要目的是为人们提供信息服务。日常生活中常用的搜索网站有百度、雅虎、搜狐等。

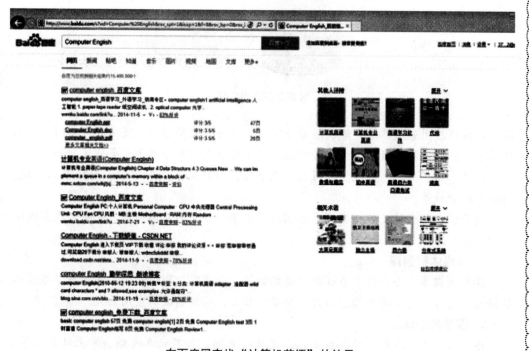

在百度网查找"计算机英语"的结果

5. 分布式数据库

分布式数据库系统适合于单位分散的部门，允许各个部门将其常用的数据存储在本地，多数处理就地完成，各地的计算机由数据通信网络相联系。

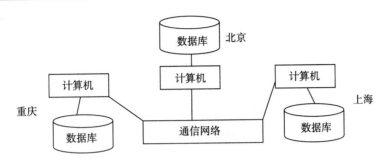

银行系统

6. 专家决策系统

专家决策系统是具有大量专门知识,并能运用这些知识解决特定领域中实际问题的计算机程序系统,如诊断系统(医疗、故障)、预报系统(天气、交通)、控制系统(战场控制)。

Chapter 3

Software

3-6　How Can TurboSquid Help You Use 3ds Max?

 **Introduction**（导读）

3ds Max is the most popular software for 3D industry professionals and TurboSquid has the largest selection of 3ds Max models available anywhere. From beginners to seasoned professionals, we have your best resource for all 3ds Max things. We will introduce it in this text.

对于三维行业专业人士来说3ds Max 是最流行的软件。TurboSquid 有庞大的3d 模型库可供选择。无论是初学者还是经验丰富的专业人士，我们都有最好的3ds Max 资源。这篇课文我们将介绍它。

 **Text**（文本）

How Can TurboSquid Help You Use 3ds Max?

TurboSquid has thousands of 3ds Max models ready for download. You can open a file from TurboSquid and start using it immediately in 3ds Max.

To find 3ds Max models at TurboSquid, type a search term in the box above all and click Search. You can download files instantly after purchase, or try some of our free models to see what TurboSquid can do for you.

TurboSquid was recently named by Autodesk, the creators of 3ds Max, as the exclusive marketplace provider of user-generated 3D models and products. Our strategic relationship allows us to support Autodesk 3ds Max models better than anyone else in the industry.

TurboSquid is home to 3ds Max products and resources of all kinds. Here is some of our top 3ds Max content.

- 3ds Max Models

  TurboSquid has the largest selection of 3ds Max models in the world.

- 3ds Max Tutorials

  Learn 3ds Max from DVDs and other training sources.

- 3ds Max Plug-ins

  TurboSquid has a variety of 3ds Max plug-ins to make your work easier and faster.

A car model of 3ds max

Training sources of 3ds max

- 3ds Max File Formats

Find out about file formats 3ds Max can import and export.

A plug-in of 3ds max

The file formats 3ds Max can import and export

Key words and expressions（重点词汇）

professional [prəˈfeʃənl] *adj.* 专业的；职业的；专业性的　*n.* 专业人士
download [ˌdaʊnˈləʊd] *v.* 将（程序、资料等）从大计算机系统输入小计算机系统；下载
term [tɜːm] *n.* 术语；期限；学期；条款
purchase [ˈpɜːtʃəs] *v.* 购买；采购；换得；依靠机械力移动
exclusive [ɪkˈskluːsɪv] *adj.* 专用的；高级的；排外的；单独的
tutorial [tjuːˈtɔːrɪəl] *n.* 个别辅导时间；教程；辅导材料；使用说明书；辅导课
plug-in [ˈplʌgɪn] *n.*（计算机软件术语）插件
format [ˈfɔːmæt] *n.*（出版物的）版式；[自]（数据安排的）形式
import [ˈɪmpɔːt] *vt.* 输入；进口
export [ˈekspɔːt] *vt. & vi.* 出口；输出

Expanded vocabulary（扩展词汇）

user-generated　*adj.* 用户生成的
strategic [strəˈtiːdʒɪk] *adj.* 战略性的；战略（上）的；有战略意义的；至关重要的
provider [prəˈvaɪdə(r)] *n.* 供应者；提供者；（尤指）维持家庭生计者

Oral practice（口语练习）

Teacher: Do you know about 3ds Max?

Student: No, I don't.
Teacher: 3ds Max is the most popular software for 3D design.
Student: What is 3D?
Teacher: 3D means three dimensions.
Student: I know. What can 3ds Max design?
Teacher: It can design many 3d modules. These models can be used in industry.
Student: I see. I believe I can study 3ds Max very well.

Exercises (练习题)

Ⅰ. Choices.

1. What is TurboSquid ready for?
   A. Download.   B. Downloads.   C. Load.   D. Loads.
2. You can open a file from TurboSquid and start using it immediately in _____.
   A. 3d Max   B. 3ds Max   C. 3d   D. 3d Maxs
3. To find 3ds Max models at TurboSquid, type a _____ in the box.
   A. search terms   B. search boxes
   C. search box   D. search term

Ⅱ. Questions and answers.

1. What do you know about the 3D design software?

2. Say something about the advantages and disadvantages of TurboSquid.

Ⅲ. Translate the following paragraph into Chinese.

Based on Autodesk Stingray, 3ds Max Interactive is a real-time engine that gives you tools to create immersive and virtual reality (VR) content. By combining the powerful 3D modeling and animation tools of 3ds Max with new tools for interactive experiences, 3ds Max Interactive aims to give you simple, efficient, and familiar workflows for creating engaging content. With 3ds Max Interactive, you can produce visually stunning games, architectural walkthroughs, or any other immersive, interactive experience.

Ⅳ. Translate the following sentences into English.

1. 你知道什么是三维吗?

2. 你知道哪些三维设计软件?

3. 3ds max 是世界上最好的三维设计软件。

4. TurboSquid 是三维模型的家。

译文

TurboSquid 如何能帮助你使用 3ds Max？

TurboSquid 拥有数以千计的可以下载的 3ds Max 模型，你能从 TurboSquid 上打开一个文件并能立刻在 3ds Max 上使用。

要在 TurboSquid 上找 3ds Max 模型，首先在对话框中键入搜索术语并点击搜索，在购买后你可以马上下载文件或者试用一下我们的免费模型来了解 TurboSquid 能为你做什么。

TurboSquid 最近被命名为 Autodesk，即 3dsMax 的创造者，作为用户生成的 3D 模型和产品的唯一市场供应商。我们的战略关系使我们能够在行业中比任何人更好地支持 Autodesk 3ds Max 模型。

TurboSquid 是 3ds Max 产品和所有类别资源的家。下面是我们顶级的 3ds Max 内容的一部分。

- 3ds Max 模型
  TurboSquid 是世界上最大的 3ds Max 模型选择库。
- 3ds Max 教程
  从 DVD 和其他的培训资源来学习 3ds Max。

3ds max 的一个汽车模型

3ds max 的培训资源

- 3ds Max 插件

  TurboSquid 拥有多种类型的 3ds Max 插件，能够使你的工作更容易、更快捷。

- 3ds Max 文件格式

  掌握 3ds Max 能够导入和导出的文件格式。

3ds max 的一个插件

3ds Max 能够导入和导出的文件格式

Chapter 3

Software

3-7 After Effects CC: What's Changed in This October 2013 Update?

Introduction（导读）

Do you know After Effects? It is the most popular software for video effects. It has many editions. We will introduce After Effects CC in this text.

你知道 After Effects 吗？它是最流行的视频特效软件。它有很多版本。在这篇课文中我们将介绍 After Effects CC。

Text（文本）

After Effects CC: What's Changed in This October 2013 Update?

The After Effects CC (12.1) update is now available to all Creative Cloud members.

Among many other changes and fixes, this update enables After Effects CC to run on Mac OS X v10.9. This full update makes it unnecessary to install the previous After Effects CC (12.0.1) patch. Unlike the After Effects CC (12.0.1) patch, the After Effects CC (12.1) update can be used to update the trial version of After Effects CC (12.0).

You can install the update through the Creative Cloud desktop application, or you can check for new updates from within any Adobe application by choosing Help > Updates. One way to check for updates is by closing all Adobe applications other than Adobe Bridge, and choosing Help > Updates in Adobe Bridge; this ensures that all processes related to Adobe video applications have been quit and can be updated safely.

Ideally, you should install the updates automatically

The icon of After Effects CC displayed at startup

through the Creative Cloud desktop application or by choosing Help > Updates, but you can also directly download the update packages from the download page for Windows or Mac OS by choosing the "Adobe After Effects CC (12.1)" update for your operating system.

We have also been working with several providers of plug-ins, codes, and hardware devices to assist them in updating their software to fix some errors and crashes. Please take this opportunity to download and install updated codes, plug-ins, and drivers from these providers, as relevant to your work.

New features in After Effects CC (12.1):
- Mask tracker
- Detail-preserving Upscale effect
- Improved performance for analysis phase for 3D Camera Tracker and Warp Stabilizer effects
- Property linking
- Improved Cinema 4D integration
- Adobe Anywhere integration and an early preview of the Media Browser panel

Key words and expressions（重点词汇）

update ['ʌpdeɪt] n. 现代化；更新；更新的行为或事例
available [ə'veɪləbl] adj. 可获得的；有空的；可购得的；能找到的
fix [fɪks] n. 困境；定位于；受操纵的事；应急措施
unnecessary [ʌn'nesəsəri] adj. 不必要的；多余的；无用的；无益的
trial ['traɪəl] n. 试验；[法] 审讯；审判；磨难；困难；[体] 选拔赛
    adj. 试验的；[法] 审讯的
install [ɪn'stɔːl] vt. 安装；安顿；安置；任命；使……正式就职
within [wɪ'ðɪn] prep. 不超过；在……的范围内；在……能达到的地方；在……内；
    在……里面
ideally [aɪ'diːəlɪ] adv. 理想地；完美地；观念上地；理论上地
automatically [ˌɔːtə'mætɪklɪ] adv. 自动地；无意识地；不自觉地；机械地
opportunity [ˌɔpə'tjuːnəti] n. 机会；适当的时机，良机；有利的环境、条件
relevant ['reləvənt] adj. 有关的；中肯的；相关联的；确切的；
    有重大意义[作用]的；实质性的
feature ['fiːtʃə(r)] n. 特征；特点；容貌；面貌；(期刊的) 特辑；故事片
upscale [ˌʌp'skeɪl] adj. 高档的；高收入的；销售对象为高收入者的
performance [pə'fɔːməns] n. 表现；表演；演技；执行
integration [ˌɪntɪ'greɪʃn] n. 整合；一体化；结合；
    (不同肤色、种族、宗教信仰等的人的) 混合
browser ['braʊzə(r)] n. 浏览程序；(用于在互联网上查阅信息的) 浏览器
panel ['pænl] n. 镶板；面；(门、墙等上面的) 嵌板；控制板

Expanded vocabulary（扩展词汇）

Mac OS 苹果操作系统
Creative Cloud 创意云
Adobe Bridge：Adobe 创意套件的控制中心
operating system 操作系统
related to 与……有联系
mask tracker 遮罩跟踪器
Detail-preserving 细节性的保护
Warp Stabilizer 扭曲稳定器
analysis phase 分析阶段

Oral practice（口语练习）

Teacher：Do you know After Effects?
Student：No, I don't.
Teacher：After Effects is the most popular software for video effects.
Student：What are video effects?
Teacher：Video effects are the gorgeous effect of the movie.
Student：I know. Is it easy to learn?
Teacher：No, it isn't. You need to study hard.
Student：I see. I believe I can study After Effects very well.

Exercises（练习题）

Ⅰ. Choices.

1. What is After Effects for?
    A. Video effects.   B. Effects.   C. Audio effects.   D. Videos effect.
2. This full update makes it _____ to install the previous After Effects CC (12.0.1) patch.
    A. necessary   B. unnecessary   C. unnecess   D. unssary
3. You can check for new updates from within any Adobe application by choosing Help > _____.
    A. update   B. dates   C. updates   D. date

Ⅱ. Questions and answers.

1. Which do you know about the software for video effects?

2. How to update After Effects CC?

Ⅲ. Translate the following paragraph into Chinese.

Adobe After Effects is Adobe's tool for video post-production which enables you to add professional looking special effects and retouches. If you use Adobe Premiere to edit your videos, then After Effects is the perfect complement to add special effects to your videos.

Ⅳ. Translate the following sentences into English.

1. 你知道 After Effects 这个软件吗？

2. 你知道哪些视频特效软件？

3. After Effects 是世界上最好的视频特效软件。

4. After Effects CC 很容易更新。

After Effects CC：2013 年 10 月的更新有什么变化？

所有的创意云成员都可以升级 After Effects CC（12.1）。

在很多其他的变化和改善中，这个更新能够让 After Effects CC 在苹果操作系统 10.9 上运行。这个完整的更新使得它没有必要安装以前 After Effects CC（12.0.1）的补丁。不像 After Effects CC（12.0.1）的补丁，After Effects CC（12.1）的更新能够被用在更新 After Effects CC（12.0）的试用版上。

你能够通过创意云桌面应用来安装更新，或者你能够使用任意的 Adobe 应用程序通过选择帮助—更新来检查新的更新。一种方法是通过关闭所有的 Adobe 应用程序来检查更新，在 Adobe Bridge 中选择帮助更新，这要保证和 Adobe 视频应用程序相关联的程序都已经被退出并能被安全地更新。

理论上，你可以通过创意云桌面应用或选择帮助—更新来自动安装更新，但是你也能从对应你的操作系统 Windows 下载页面或苹果操作系统下载页面上选择"Adobe After Effects CC（12.1）"更新来直接下载更新包。

我们也一直在与几个插件、代码和硬件设备的提供商合作，我们帮助他们更新他们的软件来修补错误和崩溃。请抓住机会从这些供应商处下载和你工作相关的代码、插件和驱动。

After Effects CC 在启动时显示的图标

After Effects CC（12.1）的新特性：

- 遮罩跟踪器
- 保护细节的放大效果
- 提高3D摄像机和扭曲稳定器分析阶段的表现
- 属性链接
- 提高4D电影的整合
- Adobe Anywhere 的整合和早期媒体浏览器面板的预览

Chapter 3

Software

3–8　What Is Macromedia Dreamweaver?

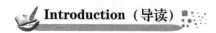 **Introduction（导读）**

Do you know Macromedia Dreamweaver? It is the most popular software for web design. It has many editions. We will introduce Macromedia Dreamweaver 4 in this text.

你知道 Macromedia Dreamweaver 吗？它是最流行的网页设计软件。它有很多版本。在这篇课文中我们将介绍 Macromedia Dreamweaver 4。

 **Text（文本）**

What Is Macromedia Dreamweaver?

Macromedia Dreamweaver 4 is a powerful Web site development software program used by professionals, as well as beginners.

The program makes it easy for designers to create visually attractive, interactive Web pages without having to know HTML or JavaScript. However, Dreamweaver 4 also enables the experienced professional to edit HTML using the new code editor.

If your department doesn't already own a copy of Dreamweaver, you may talk to your IT Department about downloading a free 30 day trial from the Macromedia website at http://www.macromedia.com. The following is the interface of the program.

1. The toolbar

The toolbar is located at the top of the page in Dreamweaver.

The toolbar

The Show Code View option displays the current page in HTML code.
The Show Code and Design Views option splits the document window, and displays the current

page in both the HTML Code view and the Design layout view.

The Show Code View

The Show code and design View

The Show Design View option displays the current page in design layout.

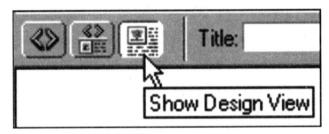
The Show design View

The title text box enables you to type the name of the document or Web page. This information will appear in the title bar of the browser.

The Title text box

Click on the Preview/Debug in Browser button, a drop-down menu appears displaying the various browsers assigned. Select the browser that you would like to preview your page in. It is important that you preview your page in Netscape and in Internet Explorer, because it is possible to see different results. After you are done previewing your page, it is important that you close the browser in order for Dreamweaver to allow you to preview your page later on.

Preview/debug in browser

2. The status bar

The status bar is located at the bottom right corner of the screen.

The status bar

3. Saving documents

The menu bar

Make sure to save your document frequently. Notice the asterisk next to the file name which indicates that you have made a change to the page without saving.

Key words and expressions（重点词汇）

powerful ['pauəfl] adj. 强大的；权力大的；（药）有效的
development [dɪ'veləpmənt] n. 发展；进化；被发展的状态；新生事物；
　　　　　　　　　　　　　　　　新产品；开发区
attractive [ə'træktɪv] adj. 迷人的；有魅力的；引人注目的；招人喜爱的
interactive [ˌɪntər'æktɪv] adj. 互动的；互相作用的；相互影响的；[计] 交互式的
edit ['edɪt] vt. 编辑；(影片、录音) 剪辑；校订；主编
department [dɪ'pɑːtmənt] n. 部门；部；系；学部；知识范围；车间
website ['websaɪt] n. [通信] 网站
toolbar ['tuːlbɑː(r)] n. （计算机）工具栏；刀杆；镗杆
view [vjuː] n. 看法；风景；视域；[建筑学] 视图
split [splɪt] vt. 分裂；分开；<俚>（迅速）离开；分担
layout ['leɪaʊt] n. 布局；安排；设计；布置图；规划图
assign [ə'saɪn] vt. 分派；选派；分配；归于；归属；把……编制
frequently ['friːkwəntli] adv. 往往；动辄；频繁地；屡次地
asterisk ['æstərɪsk] n. 星号；星状物

Expanded vocabulary（扩展词汇）

HTML 超文本标记语言
javaScript　java 脚本语言
Preview/Debug 预览/调试
drop-down 下拉
Netscape 美国 Netscape 公司（以开发 Internet 浏览器闻名）
later on 以后；嗣后；他日
status bar 状态栏

saving documents 保存文档

Oral practice（口语练习）

Teacher: Do you know Dreamweaver?
Student: No, I don't.
Teacher: It is the most popular software for Web design.
Student: What is Web design?
Teacher: Web design is designing a professional Web site.
Student: I know. Is it easy to learn?
Teacher: No, it isn't. You need to study hard.
Student: I see. I believe I can study it very well.

Exercises（练习题）

Ⅰ. Choices.

1. Dreamweaver is a powerful _____ development software program.
   A. Web site   B. site   C. Web   D. Webs site
2. The program makes it easy for designers _____ having to know HTML or JavaScript.
   A. with   B. without   C. out   D. within
3. The toolbar is located at the _____ of the page in Dreamweaver.
   A. behind   B. bottom   C. left   D. top

Ⅱ. Questions and answers.

1. Which software do you know about the Web site development?

2. How to own a copy of Dreamweaver?

Ⅲ. Translate the following paragraph into Chinese.

This new version of Adobe Dreamweaver includes new features such as Element Quick View, New editing capabilities in Live View, Live Insert to insert HTML elements directly into the Live View and a new useful Help Center.

Ⅳ. Translate the following sentences into English.

1. 你知道 Dreamweaver 这个软件吗?

2. 你知道哪些网页设计软件？

3. Dreamweaver 是世界上最好的网站设计软件。

4. Dreamweaver 很容易下载。

什么是 Macromedia Dreamweaver？

　　Macromedia 的 Dreamweaver 4 是一款被专业人士和初学者使用的、强大的网站开发软件。

　　这个程序对于设计者来说不需要掌握 HTML 或 JavaScript，就能很容易地制作形象的、吸引人的、交互式的网页。然而，Dreamweaver 4 也能让有经验的专业人士使用新的代码编辑器来编辑 HTML。

　　如果你的部门还没有一个 Dreamweaver 程序，你可以告知信息技术部门从 Macromedia 网站 http://www.macromedia.com 来下载一个免费 30 天的试用版。下面是这个程序的界面。

1. 工具栏

工具栏位于 Dreamweaver 的页面顶部。

工具栏

"显示代码视图"选项显示 HTML 代码下的当前页。

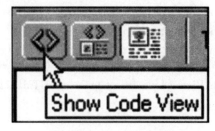

"显示代码视图"选项界面

"显示代码和设计视图"选项拆分了文档窗口，并以 HTML 和设计布局视图显示当前页面。

"显示代码和设计视图"选项界面

"显示设计视图"选项在设计视图下显示当前页面。

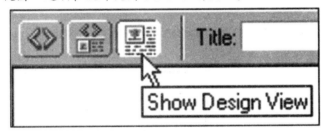

"显示设计视图"选项界面

标题文本框能够使你键入文档或网页的名字。这个信息将会在浏览器的标题栏中出现。

标题文本框

单击"在浏览器中的预览/调试"按钮，一个下拉菜单就会出现，显示出多种被指定的浏览器。选择你想预览页面的浏览器。你在 Netscape 和 Internet Explorer 中预览页面是非常重要的，因为你可能还会看到不同的结果。在你预览完页面后，最重要的是你要关掉浏览器，以便你以后能再预览此页面。

"在浏览器中的预览/调试"按钮界面

2. 状态栏

状态栏位于屏幕的右下角。

状态栏

3. 保存文件

菜单栏

一定要不断地保存文件。注意在文件名旁边的星号就表明你已经对页面做了一个改变但没有保存。

Chapter 3

Software

3-9 The Future of Adobe Fireworks

Introduction（导读）

Do you know Adobe Fireworks? It is a professional web image design, production and editing software. It has many editions. We will introduce the future of Adobe Fireworks in this text.

你知道 Adobe Fireworks 吗？它是一款专业的网页图片设计、制作与编辑软件。它有很多版本。在这篇课文中我们将介绍 Adobe Fireworks 的未来发展。

Text（文本）

The Future of Adobe Fireworks

Today at the MAX conference we announced the latest generation of our creative tools now known as CC including Photoshop CC, Dreamweaver CC, Flash Professional CC, Edge Animate CC, and many others. As you may have noticed, this announcement did not include updates to Fireworks CS6.

Over the last couple of years, there has been an increasing amount of overlap in the functionality between Fireworks and both existing and new programs like Photoshop, Illustrator, and Edge Reflow. At the same time we have shifted to focus our attention on building smaller, more modular, tools and services for specific tasks in Web design. Due to this overlap as well as our change in our product development focus, we have decided not to update Fireworks to CC and instead will focus on developing new tools to meet our customers' needs.

While we are not planning further feature development for Fireworks, we will continue to sell Fireworks CS6 as well as make it available as part of the Creative Cloud. We will provide security updates as necessary and may provide bug fixes. We plan to update Fireworks to support the next major releases of both Mac OS X and Windows. As more specific details on the next version of Windows and Mac OS X are made available, we may adjust these plans.

We understand that Fireworks has one of the most passionate communities on the Web, and that this change will be difficult to accept. Our goal in adjusting our development efforts is to build a new-generation of task focused tools that enable our customers to create great Web content—the Web Platform and Authoring team.

We appreciate all the comments following the announcement about the future of Fireworks and would like to answer a few of the common questions that are emerging:

Does Adobe care about Fireworks customers?

Absolutely—we understand that Web designers love Fireworks for it's a unique approach to page-based, stateful interaction design and rapid prototyping, and that it is an essential part of the Web design process.

Why isn't Fireworks being developed further?

Designing for the screen today is incredibly different to designing for the screen in 1998. As we considered adding new capabilities to Fireworks, we came to the conclusion that creating new, task-focused tools would better enable us to meet the future needs of Web designers and developers.

What new tools is Adobe proposing to create for Web design?

Adobe has embarked on creating a new collection of tools and services aimed at addressing the needs of today's Web designers—we've started with focusing on responsive layout, Web animation and HTML, CSS and JavaScript code editing and are delivering new Edge tools to address these use cases. We are actively working on next-generation solutions for screen design and prototyping that we hope our existing Fireworks customers will love.

The show of support for Fireworks from the community has reaffirmed our belief that Adobe should continue to deliver dedicated tools for Web designers—what follows Fireworks CS6 will be a revolutionary leap, designed from the ground up with the needs of the modern Web designers front and center. To do this we need your help. We'd love to hear about how you work, what challenges you face, where you experience the most pain in your day to day design processes.

Will Fireworks continue to be available?

Yes, Fireworks CS6 will continue to be available as part of a Creative Cloud membership.

Should I continue to use Fireworks?

Yes, if Fireworks CS6 is part of your current workflow, then there is no reason to make any change to your use of the product.

The Fireworks forum will continue to be available to CS6 users. For issues other than those related to product ownership, please post your questions on the Fireworks forum.

Is Adobe really going to fix any of the existing bugs in Fireworks?

Adobe released an update for Fireworks CS6 in mid-2013 that addressed over 25 outstanding issues, including the "File not found" issue on Mac OS 10.8 often experienced when exporting from the Image Preview dialog.

Is Adobe proposing that existing Fireworks customers switch to Photoshop?

Photoshop is a major part of the design process, but we know that Fireworks offers something unique that has made it an essential part of the Web designer's toolkit. While Photoshop is continuing to add features and workflows to support Web designers, whether or not it is a good replacement for Fireworks will depend on individual needs and preferences.

Here is the surface of starting Adobe Fireworks CS6.

The surface of starting Adobe fireworks CS6

Key words and expressions（重点词汇）

conference ['kɒnfərəns] n. 会议；讨论；（正式）讨论会；
　　　　　　　　　　　　　［工会、工党用语］（每年的）大会
　　　　　　　　　　vi. 举行或参加（系列）会议
announcement [ə'naʊnsmənt] n. 宣告；通告；布告；预告
couple ['kʌpl] n. 对；双；配偶；夫妻；＜口＞几个；两三个
　　　　　　　vt. & vi. 连在一起；连接
overlap [ˌəʊvə'læp] n. 重叠部分；覆盖物；涂盖层；［数］交叠；相交
　　　　　　　vt. 重叠；与……部分相同
　　　　　　　vi. 互搭；重叠
major ['meɪdʒə(r)] adj. 主要的；重要的；大调的；主修的（课程）
　　　　　　　n. 主修科目；大调；陆军少校
　　　　　　　vi. ＜美＞主修；专攻；［美国英语］［教育学］主修（in）；专攻
passionate ['pæʃənət] adj. 激昂的；热烈的；易怒的；易被情欲所支配的
community [kə'mjuːnəti] n. 社区；共同体；社会团体；［生态］群落
authoring ['ɔːθərɪŋ] n. 著作
approach [ə'prəʊtʃ] vt. & vi. 接近；走近；靠近

      vt. 接近；着手处理；使移近；试图贿赂（或影响、疏通）
      n. 方法；途径；接近
      vi. 靠近
incredibly [ɪnˈkredəbli] *adv.* 难以置信地；很；极为
embark [ɪmˈbɑːk] *vi.* 上飞机；上船；着手；从事
      vt. 使……上船或飞机；使从事，使着手；投资于
deliver [dɪˈlɪvə(r)] *vt.* 交付；发表；递送；使分娩
      vi. 传送；投递
reaffirm [ˌriːəˈfɜːm] *vt.* 重申；再肯定
dedicated [ˈdedɪkeɪtɪd] *adj.* 专用的；专注的；投入的；献身的
      v. 奉献（dedicate 的过去式和过去分词）
leap [liːp] *vi.* 跳；冲动地行动
      vt. 跳过；跃过；使跳跃
      n. 跳跃；飞跃；跳跃的距离
forum [ˈfɔːrəm] *n.* 论坛；讨论会；专题讨论节目；集会的公共场所；
      提供公开讨论的媒体；法庭
issues [ˈɪʃjuːz] *n.* （水等的）流出（issue 的名词复数）；出口；放出；
      （特别重要或大众关注的）问题
      v. 出版（issue 的第三人称单数）；发表；宣布；分配
ownership [ˈəʊnəʃɪp] *n.* 所有权；所有；所有制；物主身份
outstanding [aʊtˈstændɪŋ] *adj.* 杰出的；显著的；凸出的；未完成的
propose [prəˈpəʊz] *vt.* 求婚；提议；建议；打算；计划；推荐；提名
      vi. 求婚；做出计划；打算
preferences [ˈprefərənsɪz] *n.* 参数选择；较喜欢的东西（preference 的名词复数）；
      优待；偏爱的事物；最喜爱的东西

 Expanded vocabulary（扩展词汇）

prototyping [ˌprəʊtəˈtaɪpɪŋ] *n.* 原型机制造
task-focused 以任务为中心的
responsive layout 响应式布局
workflow [ˈwɜːkfləʊ] *n.* 工作流程
toolkit [ˈtuːlkɪt] *n.* 工具包；工具箱
whether or not 是否；无论

Oral practice（口语练习）

Teacher：Do you know about Fireworks?
Student：No, I don't.
Teacher：It is the most popular software for Web image design.
Student：What is Web image design?

Chapter 3　Software

Teacher: Web image design is designing images for Web site.
Student: I know. Is it easy to learn?
Teacher: No, it isn't. You need to study hard.
Student: I see. I believe I can study it very well.

Exercises（练习题）

Ⅰ. Choices.

1. Fireworks is a professional _____ software program.
   A. Web design　　　B. Web image design　　C. images　　　D. Web
2. Over the last couple of years, there has been an increasing amount of _____ in the functionality between Fireworks and both existing and new programs.
   A. over　　　　　　B. lap　　　　　　　　C. overlap　　　D. overlaps
3. While we are not planning further feature development for _____, we will continue to sell Fireworks CS6 as well as make it available as part of the Creative Cloud.
   A. Fireworks　　　　B. Firework　　　　　C. Photoshop　　D. Dreamweaver
4. We understand that Fireworks has one of the most passionate communities on the Web, and that this change will _____ to accept.
   A. be difficults　　　B. difficult　　　　　C. difficults　　　D. be difficult
5. Designing for the screen today is _____ different to designing for the screen in 1998.
   A. incredibly　　　　B. incredibe　　　　　C. incredible　　　D. incredibing

Ⅱ. Questions and answers.

1. Which software do you know about the Web image design?

2. Will Fireworks continue to be available?

3. Could you continue to use Fireworks?

4. Is Adobe proposing that existing Fireworks customers switch to Photoshop?

5. Is Adobe really going to fix any of the existing bugs in Fireworks?

Ⅲ. Translate the following paragraph into Chinese.

The complete set of tools included in Adobe Fireworks helps you design and create pretty much any element that could be used on the Web, from a simple button to a complex website layout. Plus, the new version has extended its boundaries beyond the computer Web browser, and lets you work for other devices as well, such as kiosks and smartphones.

Ⅳ. Translate the following sentences into English.

1. 你知道 Fireworks 这个软件吗?

2. 你知道哪些网页图像制作软件?

3. Fireworks 是专业的网站图像设计软件。

4. Fireworks 不再更新版本了。

5. Fireworks 在论坛上有很多爱好者。

Adobe Fireworks 的未来

今天在 MAX 发布会上我们宣布了最新一代创意工具,现在被称为 CC,它包括 Photoshop CC, Dreamweaver CC, Flash Professional CC, Edge Animate CC 和许多其他的软件。你可能已经注意到这一声明不包括 Fireworks CS6 的更新。

在过去的几年中, Fireworks 和已存在的以及新程序如 Photoshop, Illustrator 和 Edge Reflow 之间有越来越多的重叠功能。与此同时,对于网页设计的具体任务我们

Chapter 3 Software

已经转移团队注意力而专注于构建更小、更模块化的工具和服务。由于功能重叠,以及我们的产品开发重点的变化,我们决定不更新 Fireworks 到 CC,而是将重点放在开发新的工具以满足客户的需求。

虽然我们没有计划进一步开发 Fireworks 的功能,但我们将继续销售 Fireworks CS6 并使其可作为创意云的一部分。我们将提供必要的安全更新,并可能提供错误修复。我们计划更新 Fireworks 来支持 Mac OS X 和 Windows 的下一个主要版本。如果 Windows 和 Mac OS X 下一个版本的更多细节是可用的,我们可以调整这些计划。

我们知道,Fireworks 是网络上最火热的社区之一,这种变化将令人难以接受。我们重新调整的发展努力的目标是建立一个新一代的专注于任务的工具,使我们的客户创造大量的网页内容——网络平台和创作团队。

我们感谢大家在这次新版 Fireworks 发布会之后给出宝贵的评价,并愿意回答以下几个即将出现的普遍问题:

Adobe 关心 Fireworks 客户吗?

绝对的——我们了解网页设计师喜爱 Fireworks 是因为它独一无二的、基于页面的动态状态交互设计和快速成型的方法,而且它是网页设计过程中必不可少的一部分。

为什么 Fireworks 不更深入发展?

今天的屏幕设计与 1998 年的屏幕设计有着难以置信的不同。当我们考虑为 Fireworks 增加新功能的时候,我们得出的结论是创造新的以任务为重点的工具将更好地使我们能够满足网页设计师和开发人员未来的需求。

Adobe 为网页设计创造了哪些新的工具?

Adobe 已经开始创造一套新的采集工具和服务,旨在解决当今网页设计师的需要——我们开始专注于响应式布局、网络动画、HTML、CSS 和 JavaScript 代码编辑和提供新的边缘工具来解决这些使用案例所涉及的问题。我们正在积极致力于下一代屏幕设计和原型解决方案,希望我们现有的 Fireworks 客户将会喜欢。

社区对 Fireworks 的支持坚定了我们的信念,Adobe 应继续为网页设计师开发提供专用工具——这对于 Fireworks CS6 将是一个革命性的飞跃,因为 Fireworks CS6 从头到尾都是在现代高端设计师的需求下开发的。要做到这一点,我们需要你的帮助。我们很想听听你是如何工作的、你面临着什么样的挑战、你在日常设计过程中最大的痛苦在哪里。

Fireworks 将继续可用吗?

是的,Fireworks CS6 将继续作为创意云成员的一部分。

我应该继续使用 Fireworks 吗?

是的,如果 Fireworks CS6 是你目前的工作流程的一部分,那么就没有理由去改变你所用的产品。

Fireworks 论坛将继续对 CS6 用户可用。有关产品所有权以外的其他问题,请在 Fireworks 论坛发布。

Adobe 真的要修复 Fireworks 中存在的任何错误吗？

Adobe 在 2013 年年中发布了针对 Fireworks CS6 解决 25 个突出问题的更新，包括在 Mac OS 10.8 中经常出现的输出图像预览对话框"文件未找到"的问题。

Adobe 会建议现有的 Fireworks 用户转移到 Photoshop 吗？

Photoshop 是设计过程中的一个重要组成部分，但我们知道 Fireworks 提供一些独特的东西使它成为网页设计师工具包的重要组成部分。而 Photoshop 继续添加新功能和工作流程来支持网页设计师，对于 Fireworks 它是否是一个很好的替代将取决于个人的需求和偏好。

Adobe Fireworks CS6 启动界面

Chapter 3

Software

3–10　What Is QuickTime 7?

Introduction（导读）

Do you know QuickTime player? It is a professional media player software. It has many editions. We will introduce the features of QuickTime 7 in this text.

你知道 QuickTime 播放器吗？它是一款专业的媒体播放软件。它有很多版本。在这篇课文中我们将介绍 QuickTime 7 的特性。

Text（文本）

What Is QuickTime 7?

A powerful multimedia technology with a built-in media player, QuickTime lets you view Internet videos, HD film trailers and personal media in a wide range of file formats. And it lets you enjoy them with remarkably high quality.

It's a multimedia platform. For videos from your digital camera or mobile phone, a film on your Mac or PC, a media clip on a website, no matter what you're watching or where you're watching it, QuickTime technology makes it all possible.

It's a sophisticated media player. With its simple design and easy-to-use controls, QuickTime Player makes everything you watch even more enjoyable. Its clean, uncluttered interface never gets in the way of what you're watching. Want to speed through a film or slow things down? A handy slider lets you set playback from 1/2x to 3x the normal speed. And you can search through individual film frames quickly.

It's a flexible file format. QuickTime lets you do more with your digital media. With QuickTime 7 Pro, you can convert your files to different formats and record and edit your work. Third-party plug-ins extend QuickTime technology in many different directions. And QuickTime streaming solutions let you stream your media across the Internet.

The free QuickTime Player provides a host of new features.

- H. 264 video support. This state-of-the-art, standards-based codec delivers exceptional-quality videos at the lowest data rate possible, across data rates ranging from 3G to HD and beyond.

- Live resize. Playback continues smoothly as you change the size of the QuickTime Player window. (Some hardware requirements may apply.)
- Zero-configuration streaming. You no longer need to set your Internet connection speed in QuickTime Preferences. QuickTime automatically determines the best connection speed for your computer. If a connection is lost during streaming, QuickTime automatically reconnects to the server.
- Surround sound. QuickTime Player can now play up to 24 channels of audio. With QuickTime 7, your Mac, and surround speakers, you can enjoy the full effect of your surround sound movies or games.
- New and improved playback controls. Use the new A/V Controls window to adjust settings for the best viewing experience. Easily change settings including jog shuttle, playback speed, bass, treble, and balance.
- All-new content guide. The all-new QuickTime Content Guide provides the latest in entertainment on the Internet.
- Full-screen playback. Get the most out of your display by using every pixel possible. These new modes allow you to fit the content to any size screen.
- Floating controls. Full-screen mode now provides floating DVD-like controls for easy access to functions like pause, play, fast-forward, rewind, and full-screen options. Move your mouse and the full-screen controller appears on the screen for several seconds.
- Additional keyboard shortcuts. QuickTime Player now supports the same transport control keyboard shortcuts as Final Cut Pro. While viewing a movie, press J, K, or L to rewind, pause, or resume playback at variable speeds.
- Closed captioning. An option in QuickTime Player Preferences allows you to display standard EIA-608 closed captions, when they're available in your movies.

The surface of quicktime 7 playing video

Chapter 3　Software

- Time code display. QuickTime Player now allows you to switch between displaying movie time, time code, and frame count. You can also jump to a specific time code or frame number using the keyboard.
- Spotlight-friendly media. With Mac OS X v10.4 or later, you can use Spotlight to easily find your QuickTime content. Spotlight can search for movie attributes such as artist, copyright, codec, and so on.
- Screen reader compatibility. Using Voice Over, including Mac OS X v10.4 or later, visually impaired users can enjoy QuickTime Player features.

Here is the surface of QuickTime 7 playing a video.

Key words and expressions（重点词汇）

multimedia ［ˌmʌltiˈmiːdiə］ n. 多媒体
　　　　　　　　　　　　 adj. 多媒体的
trailer ［ˈtreɪlə(r)］ n. 拖车；追踪者；＜美＞拖车式活动房屋；
　　　　　　　　　　（电影或电视节目的）预告片
　　　　　　　vt. 用拖车运
　　　　　　　vi. 乘拖带式居住车旅行
remarkably ［rɪˈmɑːkəblɪ］ adv. 引人注目地；明显地；非常地
clip ［klɪp］ n. 修剪；（塑料或金属的）夹子；回纹针；剪报
　　　　　vt. & vi. 用别针别在某物上；用夹子夹在某物上
　　　　　vt. 剪；剪掉；缩短；给……剪毛（或发）
　　　　　vi. 修剪；剪；剪下报刊上的文章（或新闻、图片等）；迅速行动
sophisticated ［səˈfɪstɪkeɪtɪd］ adj. 复杂的；精致的；富有经验的；深奥微妙的
　　　　　　　　　　　　　　 v. 使变得世故；使迷惑；篡改（sophisticate 的过去分词形式）
uncluttered ［ˌʌnˈklʌtəd］ adj. 整齐的；整洁的
handy ［ˈhændi］ adj. 方便的；手巧的；手边的；附近的；便于使用的
slider ［ˈslaɪdə(r)］ n. 滑块；滑雪者；滑冰者；会滚动之物；［棒］弧度不大的曲球
playback ［ˈpleɪbæk］ n. 录音重放；录音重放装置；重放
flexible ［ˈfleksəbl］ adj. 灵活的；柔韧的；易弯曲的；易被说服的
stream ［striːm］ n. 河流；小河；川；溪；潮流；趋势；倾向；（事件等的）连续；
　　　　　　　　（财富等的）滚滚而来；流出；流注；一连串
　　　　　vt. & vi. 流；流动
　　　　　vi. 飘扬；招展；鱼贯而行；一个接一个地移动
　　　　　vt. 按能力分班（或分组）
preference ［ˈprefərəns］ n. 偏爱；优先权；偏爱的事物；（债权人）受优先偿还的权利
shuttle ［ˈʃʌtl］ n. 航天飞机；（织机的）梭子；（缝纫机的）滑梭；
　　　　　　　 短程穿梭运行的飞机（或火车、汽车）
　　　　　vt. & vi. 穿梭般来回移动
　　　　　vt. 以短程往复方式运送（货物等）

— 131 —

       vi. 以短程往复式运行
bass［beɪs］*n.* 低音歌唱家；低音乐器；[鱼]欧洲鲈鱼；[植]椴树；
      椴属树木；美洲椴木
     adj. 低音的
treble［'trebl］*adj.* 高音的；三倍的；三重的；最高声部的；尖锐刺耳的
     vt. & vi. 使成为三倍；增加两倍
     n. 三倍；最高音部；高音
     vi. 变成三倍
rewind［ˌriː'waɪnd］*v.* 重绕；倒回（影片、录音带等）
     n. 重绕；倒带器
shortcuts *n.* 捷径（shortcut 的名词复数）；近路；快捷办法；
    被切短的东西（尤指烟草）
resume［rɪ'zjuːm］*v.* 继续；重新开始；恢复职位
     n. 简历；摘要
variable［'veəriəbl］*adj.* 变化的；可变的；[数]变量的；[生]变异的
     n. 可变因素；变量；易变的东西
caption［'kæpʃn］*n.* 字幕；标题；说明文字；第三档
     vt. 给（图片、照片等）加说明文字；在（文件等）上加标题；
     在……上加字幕
standard EIA – 608 closed captions 符合 EIA – 608 标准的隐藏字幕
Spotlight［'spɒtlaɪt］*n.* Mac OS 中的快速搜索工具
compatibility［kəmˌpætə'bɪləti］*n.* 适合；互换性；通用性；和睦相处
visually［'vɪʒuəli］*adv.* 视觉上；外表上；看得见地；形象化地
impaired［ɪm'peəd］*adj.* 受损的；出毛病的；有（身体或智力）缺陷的
     v. 损害；削弱（impair 的过去式和过去分词）

Expanded vocabulary（扩展词汇）

HD film 高清电影
third-party plug-in 第三方插件
H. 264 video 高清解码视频
state-of-the-art 使用最先进技术的；体现最高水平的
standards-based 基于标准的
exceptional-quality 卓越的品质
zero-configuration 零配置
full-screen 全屏的
floating control 浮动控制
fast-forward 快进

Chapter 3　Software

Oral practice（口语练习）

Teacher: Do you know about QuickTime player?
Student: No, I don't.
Teacher: It is a professional media player software.
Student: Oh, can I use it to see videos on the Internet?
Teacher: Yes. For videos from your digital camera or mobile phone, a film on your Mac or PC, a media clip on a website, no matter what you're watching or where you're watching it, QuickTime player makes it all possible.
Student: I know. Is it easy to use?
Teacher: Yes, it is. You can use it easily.
Student: I see. I believe I can use it very well.

Exercises（练习题）

Ⅰ. Choices.

1. QuickTime is a professional _____ software program.
   A. media   B. media player   C. players   D. medias players
2. A powerful _____ technology with a built-in media player, QuickTime lets you view Internet videos, HD film trailers and personal media in a wide range of file formats.
   A. multimedia   B. multi   C. multimedia's   D. media
3. With QuickTime 7 Pro, you can convert your files to _____ and record and edit your work. Third-party plug-ins extend QuickTime technology in many different directions.
   A. formats   B. different format
   C. different formats   D. any formats
4. You no longer need to set your Internet connection speed in QuickTime _____.
   A. preference   B. Preference   C. preferencs   D. Preferences
5. With QuickTime 7, your Mac, and surround speakers, you can enjoy the full effect of your _____ movies or games.
   A. surround sounds   B. surrounds sound
   C. surround sound   D. surrounds sounds

Ⅱ. Questions and answers.

1. Which software do you know about the media player?

2. What media can play for QuickTime 7?

— 133 —

3. What features does the free QuickTime Player provide?

4. Do you know about the H. 264 video?

5. What is surround sound?

Ⅲ. Translate the following paragraph into Chinese.

The QuickTime File Format (QTFF) is designed to accommodate the many kinds of data that need to be stored in order to work with digital multimedia. The QTFF is an ideal format for the exchange of digital media between devices, applications, and operating systems, because it can be used to describe almost any media structure.

Ⅳ. Translate the following sentences into English.

1. 你知道 QuickTime 这个软件吗？

2. 你知道哪些视频播放软件？

3. QuickTime 是专业的视频播放软件。

4. QuickTime 软件使用很方便。

5. QuickTime 有很多使用者。

QuickTime 7 是什么？

QuickTime 是一种功能强大的、具有内置媒体播放器的多媒体技术，它让你观看互联网视频、高清电影预告片和各种文件格式的个人媒体。它让你享受高质量的多媒体效果。

这是一个多媒体平台——你的数码相机或手机里的视频、你的 Mac 或 PC 上的电影、一个网站上的媒体剪辑，不管你看什么或者你在哪里看，QuickTime 技术都能使其成为可能。

QuickTime 播放器是一个复杂的媒体播放器。它具有简单的设计和易于使用的控件，能让你的观映体验更加愉快。它干净、整洁的界面从来没有干扰你看什么。想快进还是慢放？一个方便的滑块可以将播放速度设置为正常速度的 1/2~3 倍。你可以快速搜索单个的电影帧。

QuickTime 播放器有一个灵活的文件格式。它让你的数字媒体功能更多。有了 QuickTime 7 Pro，你可以转换文件到不同的格式并记录和编辑你的操作。第三方插件使得 QuickTime 技术扩展到许多不同的方向。QuickTime 流媒体解决方案让你的媒体通过流出现在互联网上。

免费的 QuickTime 播放器具有一些新的特点：

- H.264 视频支持。这个最先进的、基于行业标准的解码器提供了在最低的数据速率下，以及数据速率从 3G 到高清及超越高清的卓越的视频质量。
- 实时的重新设置大小。回放能像你改变 QuickTime 播放器窗口的大小一样平滑地进行。（需要满足一些硬件要求。）
- 零配置。你不再需要用 QuickTime 偏好设置你的 Internet 连接速度。QuickTime 会自动为你的计算机确定最佳的连接速度。如果连接丢失，QuickTime 会自动重新连接服务器。
- 环绕声。QuickTime 播放器可以播放音频多达 24 个通道。使用 QuickTime 7、Mac 和环绕声音箱，你可以享受环绕声电影或游戏的全部特效。
- 新改进的播放控制。使用新的 A/V 控制窗口能调整最佳观赏体验的设置。轻松更改包括步进变速、播放速度、低音、高音和平衡的设置。
- 新内容向导。全新的 QuickTime 内容向导提供了最新的网络娱乐。
- 全屏幕播放。通过使用一切可用的像素来充分利用你的播放器。这些全新的播放模式能使你将观看内容调整到任何尺寸的屏幕下观看。
- 浮动控制。全屏幕模式现在提供浮动式的像 DVD 一样的控制，使你能轻松地访问类似暂停、播放、快进、倒带和全屏选项功能。移动鼠标，全屏控制器就会出现在屏幕上几秒钟。
- 额外的键盘快捷键。QuickTime 播放器现在支持相同的控制键盘快捷键，像

Final Cut Pro 一样，在观看一部电影时按 J、K 或 L 可以快退、暂停或以可变的速度恢复回放。
- 隐藏字幕。QuickTime 播放器偏好设置的一个选项是当电影里有隐藏字幕时可以显示符合 EIA-608 标准的隐藏字幕。
- 时间码显示。QuickTime 播放器现在可以在电影时间、时间码和帧计数之间切换显示。你还可以使用键盘跳转到特定的时间代码或帧数。
- Spotlight 友好式搜索媒体。Mac OS X v10.4 或之后的版本可使你利用 Spotlight 来轻松地找到你的 QuickTime 内容。Spotlight 可以搜索电影的属性，如艺术家、版权、编解码器等。
- 屏幕阅读器的兼容性。使用 Voice Over，包括 Mac OS X v10.4 及之后的版本，视障用户可以享受 QuickTime 播放器功能。

下面是 QuickTime 7 播放视频的界面。

QuickTime 7 播放视频的界面

Chapter 3

Software

3-11　New Features of Animate CC

 Introduction（导读）

Do you know Animate CC? It is a professional animation software. It is developed on the basis of flash. It has many editions. We will introduce the features of Animate CC in this text.

你知道 Animate CC 吗？它是一款专业的动画制作软件。它是在 Flash 基础上开发出来的。它有很多版本。在这篇课文中我们将介绍 Animate CC 的特性。

 Text（文本）

New Features of Animate CC

As scheduled, Animate CC has many exciting new features for game designers, animators, and educational content crews. Now let us get to know the new features.

- **Assets at your fingertips with Creative Cloud Libraries**

Browse and access brushes, colors, graphics, and other creative assets in libraries that are available in Animate and other Creative Cloud apps. Libraries sync to Creative Cloud, and assets can be linked so that when one is changed, you and your team can choose to update it across any projects.

- **Adobe Stock images and graphics**

The new Adobe Stock marketplace lets you find license and manage royalty-free images and vector graphics from directly within Animate. Choose from 45 million assets, save your selection to your Creative Cloud Libraries, and then drag it into your project to use. You can also license video assets directly from the Adobe Stock marketplace.

- **Adobe Typekit integration**

Access thousands of Typekit fonts from right within your font menu and apply them to your animate Web projects. Fonts from top-tier foundries are all part of your Creative Cloud membership.

- **Greater flexibility with vector art brushes**

New vector art brushes let you modify the path of a stroke after it's been drawn and scale to any resolution without loss of quality. You can also make custom brushes and import brushes created with Adobe Capture CC.

- **4K + video export**

Use custom resolution export to ensure that your videos will look great on the latest Ultra HD televisions and monitors.

- **Improved brushes and pencils**

Easily draw smooth, precise vector outlines along a curve and get faster live previews.

- **Custom resolution**

Give new life to older projects by resizing and optimizing them for any resolution, such as Hi DPI, Retina, and 4K displays.

- **Easy audio syncing**

Attach and control audio looping directly on the Timeline of your animations for perfect synchronization without having to code.

- **360° rotatable canvas**

Rotate the canvas on any pivot point while you draw, just as you would with paper and pencil, to get perfect angles and strokes.

- **Color changes made easy**

Name tagged colors so you can change one and have it automatically update across your entire composition.

- **HTML5 Canvas templates**

Easily produce rich interactive ads and other content by creating reusable HTML5 Canvas wrapper templates within Animate that can be modified with any code editor.

- **Colored onion skinning**

Easily orchestrate complex animations now that you can see multiple adjacent frames with different color and alpha values on the stage.

And also there are so much more features including Packaging of animations as OAM files, the ability to import SVG files directly onto the stage using Creative Cloud Libraries or the File Import command and more.

Here is the start surface of Animate CC 2017.

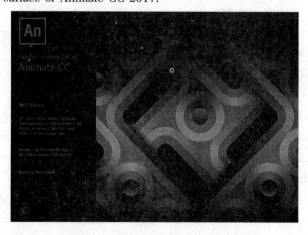

The start surface of Animate CC 2017

Chapter 3 Software

Key words and expressions（重点词汇）

asset ['æset] *n.* 资产；财产；有价值的人或物；有用的东西；优点
fingertip ['fɪŋgətɪp] *n.* 指尖；指套
graphics ['græfɪks] *n.* [测] 制图学；制图法；图表算法
sync [sɪŋk] *n. & v.* 同时；同步
license ['laɪsns] *n.* 许可证；执照；特许
            vt. 同意；发许可证
vector ['vektə(r)] *n.* 矢量；航向；[生] 带菌者；[天] 矢径
            vt. 用无线电引导；为……导航
integration [ˌɪntɪ'greɪʃn] *n.* 整合；一体化；结合；（不同肤色、种族、宗教信仰等的人的）混合
foundry ['faʊndri] *n.* 铸造厂
flexibility [ˌfleksə'bɪləti] *n.* 柔度；柔韧性；机动性；灵活性；伸缩性；可塑度
stroke [strəʊk] *n.* 中风；一击；击球；划水动作
            vt. 划掉；轻抚；轻触；敲击
            vi. 击球；作尾桨手；指挥划桨
resolution [ˌrezə'luːʃn] *n.* 分辨率；解决；决心；坚决
export ['ekspɔːt] *vt. & vi.* 出口；输出
            vt. 传播；输出（思想或活动）
            n. 输出；出口；输出[出口] 物
precise [prɪ'saɪs] *adj.* 精密的；精确的；清晰的；正规的
outline ['aʊtlaɪn] *n.* 梗概；大纲；提纲；草稿；要点；主要原则；外形；轮廓；轮廓线；轮廓画法；略图（画法）
            vt. 概述；略述；画轮廓；打草图；描略图
curve [kɜːv] *n.* 弧线；曲线；曲线状物；弯曲物；[棒] 曲线板
            vt. 使弯曲；使成曲线；使成弧形
            vi. 弯曲；弯成曲线；弯成弧形；沿曲线行进
            adj. 弯曲的；曲线形的
preview ['priːvjuː] *n.* 试映；预演；预告片；象征；预示
            vt. 预映；预先观看；概述；扼要介绍
custom ['kʌstəm] *n.* 习惯；惯例；海关；关税；经常光顾；[总称]（经常性的）顾客
            adj. （衣服等）定做的；定制的
optimize ['ɒptɪmaɪz] *vt.* 使最优化；使尽可能有效
attach [ə'tætʃ] *vt. & vi.* 附上；贴上；系
            vt. （有时不受欢迎或未受邀请）参加；把……固定；把……归因于
            vi. 附着；从属；伴随而来；联在一起（to, upon）
loop [luːp] *n.* 回路；圈；环；[医] 宫内避孕环；弯曲部分
            vt. & vi. （使）成环；（使）成圈；以环连结；使翻筋斗

synchronization [ˌsɪŋkrənaɪˈzeɪʃn] n. 同步；同一时刻；使时间互相一致；同时性
canvas [ˈkænvəs] n. 帆布；油画（布）
　　　　　　　　vt. 用帆布覆盖
　　　　　　　　adj. 帆布制的
pivot [ˈpɪvət] n. 枢轴；中心点；中枢；［物］支点；支枢；［体］回转运动
　　　　　　　vi. 在枢轴上转动；随……转移
　　　　　　　vt. 把……放在枢轴上；以……为核心；使绕枢轴旋转；由……而定
　　　　　　　adj. 枢轴的；关键的
angle [ˈæŋgl] n. 角；［比喻］（考虑、问题的）角度；观点；轮廓鲜明的突出体
　　　　　　　vt. 使形成（或弯成）角度；把……放置成一角度；调整（或对准）
　　　　　　　……的角度；使（新闻、报道等）带有倾向性
　　　　　　　vi. 垂钓；斜移；弯曲成一角度；从（某角度）报道
tagged [tægd] adj. ［医］标记的；加标记的；加了标记的
composition [ˌkɒmpəˈzɪʃn] n. 作文；作曲；创作；构图；布置；妥协；和解
templates [ˈtemplɪts] n. 模板；样板（template 的名词复数）；型板
wrapper [ˈræpə(r)] n. 包装纸；封套；封皮；（食品等的）包装材料
orchestrate [ˈɔːkɪstreɪt] vt. 精心策划；把（乐曲）编成管弦乐；和谐地安排
multiple [ˈmʌltɪpl] adj. 多重的；多个的；复杂的；多功能的
　　　　　　　　　n. ＜数＞倍数；［电工学］并联；连锁商店；下有多个分社的旅
　　　　　　　　　　行社
adjacent [əˈdʒeɪsnt] adj. 相邻的；邻近的；毗邻的；（时间上）紧接着的

Expanded vocabulary（扩展词汇）

stock [stɔk] n. 库存；树干；家畜；股份
royalty-free 免版税
top-tier 前级；顶级
4K + video 超高清的 4K 画质
Ultra HD 超高清
DPI 点每英寸（表示分辨率）
Retina [ˈretinə] n. ［解］视网膜；超高分辨率屏幕

Oral practice（口语练习）

Teacher：Do you know about Animate CC?
Student：No, I don't.
Teacher：It is the most popular software for Animation.
Student：What is Web Animation?
Teacher：It is a video technology which is formed by shooting objects and playing them continuously.
Student：I know. Is it easy to learn?

Chapter 3 Software

Teacher: No, it isn't. You need to study hard.
Student: I see. I believe I can study it very well.

Exercises (练习题)

I. Choices.

1. Browse and access brushes, colors, graphics, and other creative _____ in libraries that is available in Animate and other Creative Cloud apps.
   A. assets   B. assets   C. assetes   D. assetses

2. The new Adobe Stock marketplace lets you find, license, and manage royalty-free images and vector graphics from directly within _____.
   A. Animat   B. Animate   C. Animation   D. Animats

3. Fonts from top-tier _____ are all part of your Creative Cloud membership.
   A. foundrys   B. foundry   C. foundryes   D. foundries

4. New _____ art brushes let you modify the path of a stroke after it's been drawn and scale to any resolution without loss of quality.
   A. vectors   B. vectores   C. vector   D. vectorses

5. Use custom _____ export to ensure that your videos will look great on the latest Ultra HD televisions and monitors.
   A. resolutionary   B. resolutions   C. resolute   D. resolution

II. Questions and answers.

1. Which software do you know about Animation?

2. Does Animate CC develop on the basis of Flash?

3. What can Animate CC do?

4. What is Adobe Stock marketplace?

5. What are new vector art brushes?

Ⅲ. Translate the following paragraph into Chinese.

Studios can create enormous problems if they begin projects in a disorderly manner. By creating a thoughtful, detailed storyboard, you can clarify your ideas and get your project off on the right foot. In this course, learn how to set up a storyboard in Adobe Animate CC. Dermot O'Connor demonstrates how to set up your project, do fades and cross dissolves, create vertical pan shots, work with characters like marching robot monsters, add and edit audio, and more.

Ⅳ. Translate the following sentences into English.

1. 你知道 Animate CC 这个软件吗？

2. 你知道哪些动画制作软件？

3. Animate CC 是专业的动画制作软件。

4. Animate CC 是最新的动画制作版本。

5. Animate CC 是在 Flash 基础上开发的。

Animate CC 的新功能

Animate CC 如期而至，它为游戏设计人员、动画制作人员及教育内容编创人员推出了很多激动人心的新功能。下面我们就来了解一下这些新功能。

- **在你的指尖与创意云库上的资源**

浏览和访问画笔、颜色、图形和其他创造性的可在动画和其他创造性的云应用程

Chapter 3 Software

序库中可用的资源。库与创造性云同步,并且资源可以链接,因此当资源被改变时,你和你的团队可以选择在任何项目上更新它。

- **Adobe 库存图像和图形**

新的 Adobe 库存市场让你找到许可证且直接从动画编辑中管理免税版图像和矢量图形。从 4 500 万个资源中选择,将你的选择保存到你的创意云库中,然后将其拖到项目中使用。你也可以直接从 Adobe 库存市场授权视频资源。

- **Adobe Typekit 集成**

通过字体菜单访问数以千计的 Typekit 字体,将它们应用到你的动画网页项目中。顶级铸造字体都是你的创意云成员。

- **更大灵活性的矢量艺术画笔**

新的矢量艺术画笔让你能修改被绘制和缩放到任何分辨率的描边路径且不会丢失质量。你还可以自定义画笔和导入 Adobe Capture CC 创建的画笔。

- **4K + 视频输出**

使用自定义分辨率输出可以确保你的视频在最新超高清电视和显示器上看起来都非常棒。

- **改进的刷子和铅笔**

很容易沿曲线画出平滑、精确的矢量轮廓并得到更快的实时预览。

- **自定义分辨率**

给予老项目调整,可优化到任意分辨率,如高分辨率、视网膜和 4K 显示。

- **简单的音频同步**

无须编码,通过连接和控制音频循环来直接对时间线上的动画进行完美同步。

- **360°旋转画布**

在你绘画时通过任何支点旋转画布,得到完美的角度和轮廓,就像你用纸和铅笔一样。

- **颜色改变很容易**

你可以改变一个标记颜色的名称,并自动更新整个组成。

- **HTML5 画布模板**

容易产生丰富的互动广告和其他内容,这些内容可在 Animate 中通过创建可重复使用的、可以用任何代码编辑器修改的 HTML5 画布包装模板来创建。

- **有色洋葱皮**

既然你可以看到多个相邻帧之间的不同颜色和透明度的动画,那么制作复杂的动画就很简单了。

还有更多的功能,包括打包的 OAM 动画文件,用创意云库或文件导入命令来导入 SVG 文件直接到舞台等。

下图就是 Animation CC 2017 的启动界面。

Animation CC 2017 启动界面